Plant Tissue Culture

Padmanabh Dwivedi
Department of Botany
Arunachal University
Itanagar-791111

SCIENTIFIC® PUBLISHERS
SCIENTIFICPUBONLINE.COM

Published by
SCIENTIFIC PUBLISHERS (INDIA)

Jodhpur –

5 A, New Pali Road
P.O. Box 91
Jodhpur - 342 001 INDIA

Delhi –

4806/24, Ansari Road
Daryaganj
New Delhi - 110 002 INDIA

Print : 2024

ISBN: 978-81-7233-57-7

eISBN: 978-93-87913-72-1

Visit the Scientific Publishers (India) website at
http://www.scientificpubonline.com

Printed in India

PREFACE

Plant tissue culture is the technique of growing cells, tissues and organs in an artificial nutrient medium, under aseptic conditions. It offers novel approaches to plant production, propagation and conservation. The discipline of plant tissue culture has augmented our knowledge on fundamental botany, in the field of agriculture, horticulture, forestry, plant breeding, phytopathology etc. This book on plant tissue culture is intended for students of Undergraduates and Postgraduates of various colleges and Universities. The book embodies eleven chapters, starting with introduction, principles and basic aspects including tools and techniques. It is followed by chapters on somatic embryogenesis, micropropagation, protoplast culture and somatic hybridization, synthetic seeds, production of haploids, meristem culture and production of disease-free plants, somaclonal variation, production of stress tolerant plants, cryopreservation and biodiversity and biotechnology. The importance of biodiversity in recent years and the role of plant biotechnological approaches including techniques of plant tissue culture in sustaining the biodiversity have also been presented. The methodology of techniques has been briefly given due place in the chapters, wherever necessary. Figures, illustrations and examples are presented to elucidate the topic in appropriate fashion, and these features of the book will generate interest in the readers toward the subject. Each chapter has suggested readings and questions at the end. Further, the language used is readers-friendly.

The plant tissue culture, which is an integral part of plant biotechnology, constitutes a bulk of the course in Undergraduate and Postgraduate classes of several Universities. In recent years, vast data has been generated and compiled in the form of reference books, on this subject. There have been books published on the subject in the past which have presented the subject in elaborate form but with more emphasis on research findings. This book is an attempt to furnish students and teachers the information at one place in consolidated and precise manner, covering most of the aspects of plant tissue culture, which may serve as a great help to them. The book will also be useful for research students engaged in plant tissue culture related work, as research highlights with references have been provided.

I express my sincere thanks to Professor U.N. Dwivedi, Department of Biochemistry, University of Lucknow for kindly providing photographs related to micropropagation. His help is highly appreciated. I am grateful to my parents for giving encouragement and moral support, and to my elder brother, Dr. S.K. Dwivedi, Department of Environmental Science, B.B. Ambedkar University, Lucknow for his help in various ways during the preparation of this book. I thank the Publishers, Scientific Publishers (India), Jodhpur for taking pains and keen interest in the publication of this book. Constructive suggestions, if any, are welcomed.

Arunachal University
Itanagar 791 111

Padmanabh Dwivedi

CONTENTS

ABOUT THE BOOK

Plant Tissue Culture is an integral part of plant biotechnology, and has resulted in many frontier research findings. This book on Plant tissue culture deals with the principles and most of the techniques being employed in the tissue culture. Majority of the aspects have been covered in eleven chapters with focus on methodology, advantages and applications of the techniques. The application of plant tissue culture in sustaining the deteriorating biodiversity has also been highlighted.

ABOUT THE AUTHOR

Dr. Padmanabh Dwivedi, B.Sc. (Hons.) (BHU-Malviya Medal), M.Sc. (BHU – Gold Medal), Ph.D. (Cambridge), F.C.C.S. (Cambridge), F.C.P.S. (Cambridge), F.P.E.A., is a Lecturer in Botany, Arunachal University, Itanagar. He has outstanding academic career throughout. He has received several national and international scholarships and awards including Young Scientist Award of Council of Science and Technology, Govt. of U.P., Cambridge Nehru Scholarship, UNESCO Czech Academy of Sciences Fellowship, Overseas Research Studentship Award of the U.K., INSA Visiting Fellowship, Downing Muller Scholarship of Cambridge. Dr. Dwivedi has presented research papers in several international conferences, six being abroad. He has several publications in refereed journals of national and international repute. He is a Fellow of two prestigious international academic bodies and has sceveral other recognitions and honours to his credit. He served as a Graduate member of Institute of

Biology, London and as Overseas Officer, Downing College, Cambridge University, U.K. Dr. Dwivedi is the Principal Investigator of the DBT-sponsored major research project. He has wide Educational Administrative experience including Editor-in-Chief of the Arunachal University Research Journal. Dr. Dwivedi has visited several countries viz., U.K., Czech Republic, Germany, Russia, Slovak Republic, Hungary, Sweden.

CHAPTER - 1

INTRODUCTION, TECHNIQUES AND BASIC ASPECTS

Introduction

Plant tissue culture is the technique of growing plant cells, tissues and organs in an artificial prepared nutrient medium under aseptic conditions. It has benefited and advanced the knowledge of fundamental botany, specially in the field of agriculture, horticulture, plant breeding, forestry, somatic cell hybridization, phytopathology and industrial production of plant metabolites, etc. Plant tissue culture aims to induce plant organs, or portions of them, to proliferate cells, each one of which may be regenerated into whole plants. These plants are replicas of the parent plant. Using this technique it is possible to cultivate cells in large number, or as clones from single cells. Also one can grow whole plant from isolated meristems and induce callus or even single cell to develop into complete plant, either by organogensis or directly by embryogenesis *in vitro*. Tissue culture includes culturing of tissue parts, cell aggregates and individual cells. Cell lines with desired traits obtained from plants through explant tissue or via genetic engineering culture technique can be cultured, maintained and conserved through tissue culture technique. Plant cultures have been widely employed as an experimental technique for studying organogenesis and a method for rapid clonal propagation of genetically uniform plants. The techniques are proving to be a rich source of variability with huge potential in crop improvement.

When a part of plant organ or tissue is put on a culture medium and maintained under optimum growth conditions, the meristematic cells de-differentiate and proliferate to form a mass of cells called **callus**. Under the influence of growth regulators (auxin/cytokinin) the undifferentiated cells can be induced to undergo morphogenesis into roots, shoots or both. The individual plantlets can thus grow into mature plants.

The starting material for callus production may be a small section of any part of the plant - generally called the **explant**. Root, shoot, leaf, fruit tissue, petals, hypocotyls or germinating seeds may be used as explants. These would result in calluses of diploid cells, when inoculated on suitable nutrient medium. To obtain haploid calluses, the starting material should be ovules, immature pollen or whole young anthers. The calluses formed from these cells may be unorganized ones or may form miniature embryo-like structures that regenerate directly into complete plantlets. These latter structures are called as **embryoids**. To regenerate cells that are single or are part of a callus, appropriate auxin and cytokinin are added to the growth medium. When an explant is inoculated on a suitable nutrient medium, the non-dividing quiescent cells first undergo certain changes to the meristematic state leading to the formation of undifferentiated cells (callus), the process is called **de-differentiation**. The component cells of callus have the tendency to produce a whole plant, a phenomenon described as **re-differentiation**. These two phenomena - de-differentiation and re-differentiation are inherent and called **cellular totipotency.**

The production of haploid through tissue culture from anthers or isolated microspores, and of protoplasts from higher plant cells have served as the basic tools for genetic engineering and somatic hybridization. Tissue culture technology helps to propagate plants of economic importance such as orchids and other ornamental plants in larger number through meristem culture or by other *in vitro* methods. This provides disease-free plantlets.

In the plant breeding programmes, embryo, ovary and ovule culture as well as *in vitro* pollination have been employed to overcome morphological and physiological sterility and incompatibility. One of the most important developments in the field of plant tissue culture during recent years is the isolation, culture and fusion technology that have their special importance in studies related to plant improvement by cell modification and somatic hybridization. Plant tissue culture technology is a boon in the studies of biosynthesis of secondary metabolites and provides an efficient means of producing economically important crop plant parts.

Historical Account

Important developments and pioneering work are given as under:

1878:	Vochting studied plant polarity and tried to grow small lumps of tissue *in vitro*. Demonstrated that polarity was present even in the plant cell itself.
1893:	Rechinger placed segments of poplar stem and pieces of beet and candeloin roots on a wet filter paper, and observed callus formation.
1902:	Haberlandt tried to obtain plant tissues from the isolated mesophyll cells of *Lamium* on artificial Knop's medium *in vitro*. But in vain, however, he pointed out towards a new technology - plant tissue culture.
	Knop and Prdifer developed synthetic media.
1925:	The utility of embryo culture technology became apparent when Laibach (1925, 1929) reared hybrid

	embryo (*Linum perenne* × *L. austriacum*) from non-viable seeds
1934:	Successful culture of tomato roots was demonstrated by White.
1937:	Using a synthetic medium containing inorganic salts, pyridoxine, thiamin, nicotinic acid, yeast extract and sucrose, White could maintain tomato root culture for almost 30 years.
1941:	Use of coconut milk to meet nutritional requirements of younger embryo was demonstrated by Van Overbeek *et al.*, for *Datura* embryos, and its stimulating effect was demonstrated.
1944:	First *in vitro* cultures of tobacco were used to study adventitious root formation by Skoog.
1952:	Virus-free dahlias were obtained by Morel & Martin using meristem culture.
	Philip White & R. Gautheret - Pioneers and founders of methods for cultivating excised tomato roots, root tips, root meristem and callus of cambial origin.
1953:	Isolation and culture of single cell was impossible until Muir devised the technology of agitating callus tissue in liquid medium (on a shaker), through paper-raft nurse tehnique.
1955:	Discovery of kinetin, a cell division hormone, by Miller *et al.*
1957:	Skoog & Miller obtained roots and stems from callus treated with auxin and kinetin. Morel used gibberellin for induction of proliferation of meristems and their differentiation into whole plants.
1958:	Regeneration of somatic embryos *in vitro* from the nucellus of citrus ovules, demonstrated by Maheshwari & Rangaswamy.
	F.C. Steward showed that it is possible to produce

	a complete plant from a single cell. He isolated cells of carrot root, which grew in a medium containing coconut milk and found that these isolated cells began to grow. Such embryo can develop into normal mature carrot plant, with normal roots, stalks, flowers and seeds. Vasil & Hilderbrandt (1965) used this technology to raise plant from single cell culture of *Nicotiana*.
1959:	Publication of first extensive handbook on plant tissue culture by Gautheret.
1960:	Cocking showed enzymatic degradation of cell walls to get large number of protoplasts. Cell wall acts as a barrier in plant protoplast culture. Role of cellulases and pectinases in dissolving cell wall in a buffer solution at certain pH, and isolation and protoplast culture were developed by Cocking.
1962:	Murashige & Skoog medium developed.
1964:	First haploid Datura plants obtained from pollen grains by Sipra Guha & Maheshwari.
1967:	Haploid plants obtained from tobacco pollen grains by Bourgin & Nitsch.
1969:	First successful isolation of protoplasts from a suspension culture of *Hapopappus gracilis* by Erikson & Jonassen.
1970:	Selection of biochemical mutants *in vitro* by Carlson; First achievement of protoplast fusion by Power *et al.*
1971:	First plant regenerated from protoplasts by Tekebe *et al.*
1972:	Interspecific hybridization through protoplast fusion in two *Nicotiana* species by Carlson *et al.*
1973:	Cytokinin found potential of breaking dormancy in excised capitulum explants of *Gerbera* by Pierik *et al.*

1974:	Discovery of Ti-plasmid by Larebeke *et al.*
	Biotransformation discovered in plant tissue cultures
	Methods to double the chromosome number in microspores of *Datura* and *Nicotiana* were given by Nitsch
1976:	Shoot initiation from cryopreserved shoot apices of carnation demonstrated by Seibert
1978:	Somatic hybridization of tomato and potato by Melchers *et al.*
1981:	Introduction of term 'Somaclonal Variation' by Larkin & Scowcroft
1982:	Protoplasts able to incorporate naked DNA, as shown by Krens *et al.*
1983:	Production of transformed tobacco plants following single cell transformation or gene insertion, shown by Chilton.
1984:	Vasil emphasized for the development of cell culture and somatic genetics
	Transformation of plant cells with plasmid DNA by Paszkowski *et al.*
1985:	Somatic hybrid in tobacco mediated by electrofusion, shown by Kohn *et al.*

Important Tissue Culture Researches in India

In India, the significance of *in vitro* culture technology was realized by Prof. P. Maheshwari. The initial extensive work on tissues like ovary and ovule laid the foundation of several achievements in plant cell, tissue and organ culture in India. During 1956 to 1966, attempts were initiated to culture the pollinated ovaries of *Allium cepa, Foeniculum vulgare, Hyoscymus niger, Iberis amara,* etc. Success with the culture of pollinated ovaries made it possible to culture ovaries and ovules at even earlier stages. Cultured ovaries from pollinated flowers of *Zephyranthes* developed seeds with fully differentiated embryos. The important findings from India are given as under:

— Occurrence of polyembryony due to cleavage and/or budding of zygotic embryos, on a culture medium. It is useful in the multiplication of plants producing a limited number of seeds.

— The method of ovary and ovule culture is good for rearing hybrid seeds. *In vitro* culture of ovary/ovule gives best site for artificial pollination (Shivanna, 1965). In *Gossypium hirsutum*, Joshi (1962) grew ovules with 12-celled proembryo and 500 endosperm nuclei. The hybrid ovules of the cross between *G. arboreum* and *G. hirsutum* could be grown to maturity and viable seeds were obtained.

— Embryogenesis from nuclei of unfertilized ovules of *Citrus sinensis* and *C. aurantifolia* is reported by Mitra & Chaturvedi (1972). The embryo arose from nuclear cells or the callus. It was demonstrated that the pollination/fertilization stimulus is not essential. Nucellus culture has been exploited for producing disease-free and genetically uniform plants in citrus (Mitra & Chaturvedi, 1972).

— The organogenic potentiality of endosperm has been demonstrated by Johri and Bhojwani (1965) in *Exocarpus cupressiformis*.

— Khuspe *et al.* developed methods for *in vitro* growth of immature hybrid embryos of *Carica papaya* × *C. cauliflora* and development of F_1 plants. The method of embryo culture has given impetus to 'embryo rescue' method which has its application in some intergeneric and interspecific crosses. The work of Johri & Bajaj (1965), Guha & Johri (1966) and Bhojwani & Johri (1970) in plants like *Allium cepa, Dendropthoe sp.,*

Scurrula sp. etc have been pioneering in embryo culture.

— The work on haploids provided a novel experimental system for speeding up of homozygosity, analyses of genetic combinations, and detection of mutations. In cultured anthers of *D. innoxia* (with uni- or bi-nucleate pollen), numerous embryoids grew on Nitsch-medium. The embryoids germinated to form plantlets (Guha and Maheshwari, 1966).

— Srivastava & Johri (1988) reviewed Indian contribution to morphogenesis and plant tissue culture studies (Indian Review of Life Sciences, 8 : 249). A.F. Mascarenhas (1993) has compiled a Hand Book of Plant Tissue Culture published by ICAR, New Delhi.

Plant Tissue Culture : Principles

The most common culture in plant tissue is *callus*, which is wound tissue composed of undifferentiated, vacuolated and unorganized cells.

(i) Callus Culture

For raising the callus tissues, a tissue culturist must have clear understanding of basic principles. All types of organs like shoot apex, bud, leaf, mesophyll cells, epidermis, cambium, anthers, pollen, fruit etc., when inoculated on a suitable medium under aseptic laboratory conditions are able to differentiate and multiply - callus is induced. This results into the formation of an amorphous mass of cells known as callus, which can be induced to re-differentiate on appropriate medium to develop embryoids which directly develop into plantlets, eventually giving rise to a whole viable plant. Whereas in nature wound and pathogens induce callus tissue formation in intact plants, in tissue culture, callus formation is artificially induced in explants grown in micro-environments.

The term **clone** (derived from Greek work klon, meaning a twig suitable for plant propagation) was suggested by Webber (USA) to explain those plants which were obtained by a sexual reproduction. Also means an organism obtained from a single cell through mitotic divisions. Formation of callus is the result of expansion and cell division of cells of explants. Calluses can vary in appearance and texture; generally yellow or white in colour, or often pigmented. Secretion of phenolic compounds gives brown colour to the callus. By doing repeated sub-culturing of callus, it is able to grow on a culture medium which is devoid of growth hormones. An exogenous supply of regulators is often recommended to induce callus on an explant:

(i) auxin only required (specially monocotyledons)

(ii) cytokinin alone needed

(iii) both auxin and cytokinin required

Besides growth regulators, genotype, glucose (2-4%), coconut milk, light, nutrient composition, high temperature (22-28°C) govern callus formation.

(ii) Organ Culture

A particular organ is isolated and cultured under laboratory conditions in a defined medium where they retain their characteristic structures and other features and continue to grow as usual. In organ culture, organs are not induced to form callus, therefore, it differs from the callus culture where the organization of the intact tissues is lost. This technique provides an experimental system to define the nutrients and growth factors that are usually received by the organ from other parts of the plant body and from surrounding environment.

Thus, organ culture offers an excellent experimental system to study:

(i) nutritional requirements of the organ (explant)

(ii) environmental effects

(iii) metabolite biosynthesis

(iv) development of new varieties

- Organ culture
  - Vegetative organs — root culture, leaf culture and shoot tip culture
  - Reproductive organs — Complete flower culture, isolated ovary culture, isolated ovule and embryo culture, pollen mother cell culture, seed and fruit culture

Root Culture

It refers to culture of excised radicles of aseptically germinated seeds. Root has been cultured successfully and induced to regenerate under *in vitro* conditions. Genetic stability associated with growing tips offers roots as a suitable material for large-scale micropropagation of desired plants. There are appreciable amounts of successful reports of root culture (Table 1.1). Root culture, besides offering a good material for micropropagation, is also used for production of pharmaceutically important secondary metabolites of high commercial value (in case of hairy root culture). This method also enables us to understand the process of growth and differentiation in roots besides the physiological requirements of these processes.

Table 1.1: Examples of Root culture

Plants	Explant Used	Response Observed
Albizia lebbeck	Roots	Plantlets
Dalbergia sissoo	Roots	Shoot buds
Oryza sativa	Roots	Plantlets
Vicia faba	Radicle	Rooted shoots/ plantlets

Stem Culture

The stem segments have been the most frequent explant used for *in vitro* studies (Table 1.2). These explants could be induced for direct or indirect regeneration. The callus cultures obtained from stem segment have often been looked upon as a source of somaclonal variations and a rich nutrient for secondary metabolite analysis. The ultimate success with stem segment culture depends upon the media composition, physiological status of the explant and the *in vitro* environment.

Table 1.2. Examples of Stem culture

Plants	Explant Used	Response Observed
Antirhinum majus	Shoots	Callus
Cajanus cajan	Epicotyl segment	Multiple shoots
Citrus grandis	Stem	Plantlets
Eucalyptus sp.	Stem	Multiple shoots with roots
Solanum tuberosum	Stem	Plantlets
Psidium guajava	Nodal segment	Multiple shoots with roots

Shoot tip / meristem culture

Shoot tips have got apical meristem and thus possess inexhaustible degree of regeneration potential. It refers to culture of terminal (about 0.1 - 1.0 mm) part of shoot having meristem together with primordial and developing leaves and adjoining stem tissue. Also called as 'mericloning'. Clonal propagation of woody plants, ornamentals and rare species has been possible through this method (Table 1.3). It is useful in developing virus - free plants and micropropagation and storage of genetic resources. It also provides a suitable material for biochemical, physiological and genetical studies.

Table 1.3. Examples of Shoot tip/meristem culture

Plants	Explant Used	Response Observed
Allium sativum	Shoot buds	Plantlets
Artocarpus heterophyllus	Shoot tips	Multiple shoots
Musa sp.	Shoot tips	Plantlets
Picrorhiza kurroa	Shoot tips	Plantlets
Saccharum officinanum	Infected shoot (smut)	Plantlets
Morus indica	Shoot tips	Embryoids

Leaf culture

In leaf, only a few layers of cells can respond to *in vitro* conditions and give a great degree of totipotency. The apical cells get activated and undergo morphogenic changes to form a plant. Leaf cultures have also been used for callus induction and secondary metabolite synthesis. Here, excised young leaf primordia are cultured (Table 1.4).

Table 1.4. Examples of Leaf culture

Plants	Explant Used	Response Observed
Azadirachta indica	Leaf	Shoot differentiation
Citrus sp.	Leaf	Plantlets
Ficus religiosa	Leaf	Plantlets
Triticum aestivum	Leaf	Plantlets
Zea mays	Leaf	Somatic embryoids
Rauwolfia serpentina	Leaf	Somatic embryoids

Flower Culture

Culture of flower primordia or flower bud is the basis for flower culture. Flower culture is useful in studies concerned with flower and fruit development, sex differentiation etc. eg. *Arachis hypogea, Allium cepa, Tagetes sp.*

Ovary Culture

It refers to culture of ovaries excised from pollinated or non - pollinated flowers. It is useful in obtaining hybrids between self-incompatible species and induction of polyembryony. Through ovule culture it is possible to obtain seedlings from crosses that are unsuccessful.

Anther and Pollen culture

This method is useful in developing haploid plants to get homozygous diploids. This technique finds its role in plant breeding to maintain male sterile lines. It is discussed in detail in the chapter on haploid production.

Plant Tissue Culture: Requirements

The setting up of a tissue culture laboratory needs proper planning. Usually, the available laboratory space has to be divided into distinct laboratory areas: Media preparation room/preparatory area, inoculation area, environmentally controlled culture room, analytical room and acclimatization room.

Media Preparation

Requirements include

(i) Laboratory tables or benches, chairs, hot plate and magnetic stirrer, analytical loading single pan balance (electronic balance) with precision of ± 0.001 g, weighing range 0.1 mg - 180 g, refrigerator and freezer, Double distillation water plant (Fig 1.1) and storage system, glassware, gas outlet, microwave oven, digital pH meter, autoclave, etc.

(ii) Cabinet/shelves- For storing glasswares, plastic wares, chemicals, plugs and appliances required for media preparation. Items include culture tubes/ conical flasks, Petri-dishes, measuring cylinders of variable capacity, cotton for plugs (absorbant and non-absorbant)/plastic caps (autoclavable), volumetric flasks, beakers, reagent bottles, pipettes, vacuum filtration system and glass rods.

Fig. 1.1. Double distillation water plant

Culture Media

Requirements include

Detergents or wetting agents like Tween-20, 70% alcohol and absolute alcohol, distilled water, stock solution of nutrients of tissue culture media, sucrose, agar, sterile culture vessels, chemicals of analytical grades (inorganic, organic salts, vitamins, acetic acids, growth hormones/ hormones, activated charcoal), coconut, milk, yeast extract, malt extract, casein hydrolysate and extracts of potato, carrot and tomato. Vitamins and hormonal solutions are sterilized by filtering through a microfilter of pore size 0.22 - 0.45 μm. Paper/weighing boats, magnetic stirrers, gloves, mop, scoop, wastebasket, serviceline for gas, water, electricity, vacuum pump and generator.

A culture medium has the following components:

(i) inorganic nutrients (ii) organic nutrients (iii) growth hormones, and (iv) agar, if needed. The inorganic nutrients include C, H, O, macronutrients like N, P, S, Ca, K, Mg and micronutrients like Fe, Mn, Cu, Zn, B, Mb. The organic sources are added either as (i) nitrogen source (amino acids/vitamins) or as (ii) carbon source (hexose sugars and or other carbohydrates). However, in an undefined nutrient medium, caesin hydrolysate, malt extract, water melon extract, coconut milk, yeast extract, etc are used as C and N source. The various growth hormones used are (Fig. 1.2): (i) cytokinins - commonly used are kinetin, 6-BPA (6-benzyl aminopurine) and zeatin. (ii) auxins - commonly used ones are NAA (napthaleneacetic acid), IBA (indole butyric acid), 2,4-D (2,4-dichlorophenoxyacetic acid), p-CPA (para-chlorophenoxyacetic acid), NOA (napthoxyacetic acid), and (iii) gibberellins - GA3 is the most commonly used. Agar is used at a concentration of 0.8% to 1.5%.

Some of the nutrient media used are MS medium (Murashige & Skoog, 1962), ER medium (Eriksson, 1965), B5 medium (Gamborg *et al.*, 1968), but most commonly used is the MS medium. Selection of nutrient media for a specific plant requires a series of experiments involving 'trails & errors' strategy, wherein firstly a known medium is used, then experimention with the variable factors (ratio of hormones) is done, which leads to a new suitable medium.

The hormones are tried at different combinations involving 0, 0.5, 1.0, 1.5, 2, 2.5, 5, 10 μmol l^{-1}. Likewise, different concentration of salts (½ or ¼ of MS salts) and sucrose (2 - 6%) are tried and standardised with the best combination of growth regulators. Table 1.5 gives the detailed composition of MS basal medium. The nutritional composition of White's medium and B5 medium is also given (Table 1.6 & 1.7).

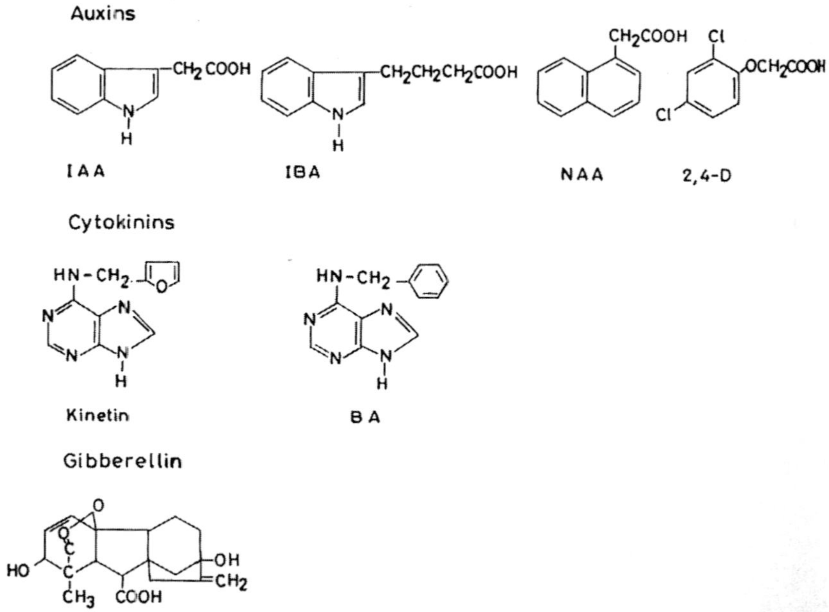

Fig. 1.2. Chemical structure of growth hormones

Normally stock solutions are prepared for different medium constituents like salts, vitamins and hormones. Dry powder media are also available and these can be dissolved in distilled water (50% less than that of final volume). The final volume is made up and agar is then added together with other constituents, pH adjusted to desired level and slightly heated so as to dissolve the agar - the gelling agent. The medium is then distributed to desired culture vessels, plugged with non- absorbant cotton and covered tightly with aluminium foil or brown paper. The flasks are then sterilized by autoclaving. Only double distilled water is recommended for making nutrient media.

<u>Aseptic transfer chamber area</u> (Inoculation room)

Laminar air flow cabinets to provide constant flow of air across the working area, dissecting and stereo zoom microscopes and lens magnifier, micro-dissecting scissors, scalpel, needles, inoculating loops, gas outlet, vacuum facility, tissue paper, sterilizer and pipette dispenser.

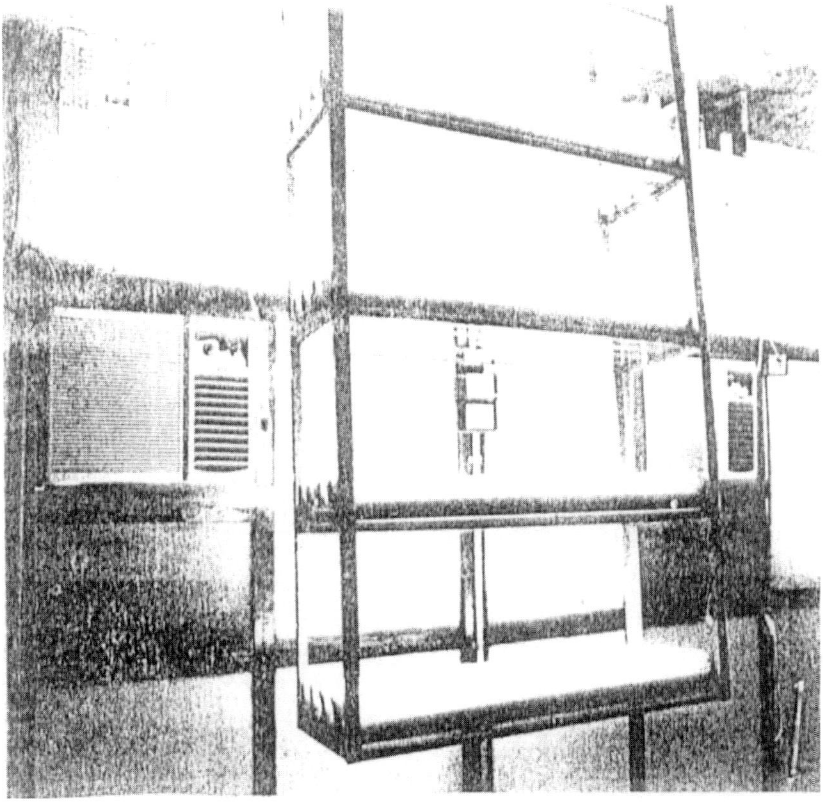

Fig. 1.3. Culture Rack with light arrangements, and Air conditioner in culture room

Environmentally controlled culture room

The requirements include

Racks with light arrangements (Fig. 1.3) and timers, and controlled temperature (25 ± 2°C) maintained with air cooler or by window air conditioners (Fig 1.3) fitted with temperature indicators; Incubators having dark and light/dark photoperiod with controlled temperature; Lux-meter, Rotating shakers; Air cutain. Cultures usually incubated at 25 ± 2°C under 16:8 light:dark photo-period, and 60-70% relative humidity. The source of light for the cultures in racks is made available with cool-day-light (fluorescent tube lights of 40 W, 3000 lux).

Table 1.5. Stock solutions for the MS basal medium

Components	Amounts taken (mg l^{-1})
Stock Solution I	
$MgSO_4.7H_20$	7400
KH_2PO_4	3400
KNO_3	38000
NH_4NO_3	33000
$CaCl_2.2H_2O$	8800
Stock Solution II	
H_3BO_3	1240
$MnSO_4.4H_2O$	4460
$ZnSO_4.7H_2O$	1720
$Na_2MoO_4.2H_2O$	50
$CuSO_4.5H_2O$	5
$CoCl_2.6H_2O$	5
Stock Solution III	
$FeSO_4.7H_2O$	5560

$Na_2.EDTA.2H_2O$	7460
Stock Solution IV	
Inositol	20000
Thiamine-HCl	100
Pyridoxine-HCl	100
Nicotinic acid	100
Glycine	400

Note: (1) In order to prepare 1 litre of nutrient medium take 50 ml of stock solution I, 5 ml each of stock solutions II - IV.

(2) The naturally occurring auxin , IAA is added in a concentration of 0.01 - 10.0 mg l^{-1} from a stock solution. The synthetic and relatively more active auxins (IBA, NAA & 2,4-D) are used at concentrations of 0.001 - 10 mg l^{-1}. Kinetin is used at a concentration of 0.04 - 10.0 mg l^{-1}.

Analytical Room

It comprises of following facilities

Inverted microscope for bright field, dark field with compensating wide angle eye pieces, colorimeter for estimation of chlorophyll, starch, nucleic acid, phenols, oxidizing enzymes etc. Low speed centrifuge with continuous variable electronic speed control, chemical reagent racks, gas outlet.

Acclimatization Room

The hardening chamber requires high illumination (4000 - 10,000 lux) and high humidity (90 - 100%, through mist and fog systems). Humidity is required for conditioning tissue culture plants after being taken out from rooting media and transferred to pots under greenhouse.

Table 1.6. Nutritional components of White's medium (White, 1953)

Components	Amounts (mg l^{-1})
Macronutrients	
$MgSO_4.7H_2O$	750
$NaH_2PO_4.H_2O$	19
KNO_3	80
Micronutrients	
H_3BO_3	1.5
$MnSO_4.4H_2O$	5
$ZnSO_4.7H_2O$	3
$CuSO_4.5H_2O$	0.01
KI	0.75
Sucrose (g)	20
Organic constituents	
Vitamins	
Thiamine-HCl	0.01
Pyridoxine-HCl	0.01
Nicotinic acid	0.05
Others	
Glycine	3
pH	5.8

Aseptic Techniques

It is advised to have complete aseptic condition around the culture equipments which prevents contamination of the culture medium. Various sources of contamination of the medium and the subsequent methods to check them have been proposed:

(1) Microorganisms may be present in nutrient medium at the time of its preparation. These microoragnisms can be destroyed by proper plugging and autoclaving the culture flasks. Medium is sterilized completely by maintaining it at 121°C for about 20 min at pressure of 1.0 - 1.5 kg cm^{-2} in autoclave.

(2) The explants may carry microorganisms hence these are surface sterilized by mercuric chloride (0.1 - 1.0%) or by sodium hypochlorite solution (0.5 - 5%) for 5 - 30 min.

(3) Precautions are taken to prevent the entry of microbes when the plug of a culture is removed during transfer of the plant material to the medium or from one medium to another. The inoculation chamber (Laminar air cabinet) may be sterilized by uv radiation.

(4) Correct pH of medium is important. Highly alkaline or acidic pH affects the nutrient uptake in cultured tissues, so tissue culture medium is adjusted to a pH of 5.6 - 6.0 before autoclaving.

(5) Semi-solid and liquid media are most commonly used for growing plant cells. A high concentration of gelling agent (agar, gelatin, silica gel) makes the medium very hard and thus decreases the nutrient uptake by the tissues.

Sterilization of Plant Tissue

Tissues (explants) are washed in a weak detergent solution and rinsed many times with double distilled water prior to sterilization. Some woody tissues buds, twigs are cleaned by immersing them briefly in a 70% ethanol solution.

— NaOCl (0.025 - 0.25%)

— CaOCl - less damage to plant tissues but tends to precipitate out of solution

— H_2O_2 (3.0% - 10%) - Easily removed from tissues

— $MgCl_2$ (0.1 - 1.0%), bromine water (1.0 - 2.0%), $AgNO_3$ (1.0%).

Non-living materials (articles) are steam-sterilized in an autoclave at a pressure of 1.0-1.5 kg cm^{-2} and a temperature 121°C for 15 min. Thermolabile compounds are sterilized at room temperature using a bacterial filter and then added to the sterilized medium aseptically.

Table 1.7. Nutritional composition of B5 medium

Components	Amounts (mg l^{-1})
Macronutrients	
$MgSO_4.7H_2O$	250
$NaH_2PO_4.H_2O$	150
KNO_3	2500
$CaCl_2.2H_2O$	150
$(NH_4)_2SO_4$	134
Micronutrients	
H_3BO_3	3
$MnSO_4.4H_2O$	10
$ZnSO_4.7H_2O$	2
$Na_2MoO_4.2H_2O$	0.25
$CuSO_4.5H_2O$	0.025
$CoCl_2.6H_2O$	0.025
KI	0.75
Sucrose (g)	20
Organic constituents	
Vitamins	
Thiamine-HCl	10
Pyridoxine-HCl	1
Nicotinic acid	1
Myoinositol	100
pH	5.5

Miscellaneous Points for Consideration

1. While preparing a nutrient medium, following kinds of interaction are to be borne in mind:

(i) Light/dark - auxin: Indoleacetic acid breaks down more in light than in dark.

(ii) Growth regulator/temperature: Changes in organ formation can be based on the availability of auxin and /or cytokinin (i.e., endogenous or exogenous). The endogenous auxin and/or cytokinin concentrations are often influenced by temperature.

(iii) Genotype/growth regulator: The requirement of a growth regulator (high or low concentration) for growth and development is dependent on genotypes.

(iv) Sugar/light: Sometimes light and sugar can replace each other.

2. The choice of nutrient media to be used for culturing the explant depends on many factors:

(i) The experimental plant: Some species, for instance, are very sensitive to salt while others can tolerate a high range of salt concentration. Similarly, some species react to addition of Vit B_1 and others do not. Also, the requirements of growth regulators vary from species to species.

(ii) Age of plant: Juvenile tissues can regenerate roots in absence of auxin but the adult tissues need the presence of auxin.

(iii) Age of the organ (explant): Actively dividing cells (young organs) have different hormonal requirements.

TOTIPOTENCY

Totipotency is the capacity of living plant cells to produce different kinds of plant cells. A totipotent cell under suitable conditions can reproduce an entire plant since it has all the required information for inheritance of traits. Morgan (1901) coined the term 'totipotency' to denote the capacity of cell to develop into an organism by regeneration. Steward *et al.* showed this process in carrot cultures. Small pieces of mature carrot root cells were grown in a liquid medium supplemented with coconut milk. These cultures were shaken gently which freed all the cells into the medium. When these cells were transferred to tube having a semi-solid medium, they produced whole plants that flowered and set seeds. The demonstration of cellular totipotency in a real way was carried out by Muir *et al.* (1958) who showed the formation of a plant body from an actively growing callus. Later on, Vasil and Hilderbrandt (1965) raised the whole plants of tobacco by first growing isolated single cells into masses of tissue and then transferring them to flasks. Based on these experiments, it could be theoretically concluded that every living plant cell, irrespective of its age and location, is totipotent.

This potentiality has been exploited through the culture of protoplasts, cells, tissues and organs *in vitro*. In cultured materials it has been possible to study such processes as differentiation of parenchyma cell into tracheid (cytodifferentiation), organ formation (organogenesis) and somatic embryo formation (somatic embryogenesis). The study of fundamental process of differentiation, the capacity of cell to form organs and embryo can be exploited to regenerate plantlets for clonal (clone is vegetatively obtained group of plants from an individual) propagation.

Organogenesis

Organogenesis (development of organs) starts with stimulation caused by chemicals of medium, substances carried over from original explants and endogenous compounds produced by culture (Thomas & Davy, 1975). This system is commonly produced in callus cultures, but can be produced directly from the explant. The earliest reports on controlled organogenesis *in vitro* were by White (1939) who obtained shoots on callus of a tobacco hybrid, and by Nobecourt (1939), who observed root formation in carrot callus. Studies with tobacco callus demonstrated that a high ratio of auxin to cytokinin in the medium favoured root formation, a reverse ratio having high cytokinin favours shoot formation, whereas intermediate ratios promoted callus formation. Skoog (1944) for the first time showed that organogenesis could be chemically controlled. He observed rhizogenesis and caulogenesis (shoot inhibition) after addition of auxin to the medium. Later on, Skoog & Miller (1957) demonstrated that a high ratio of auxin: cytokinin stimulated formation of root in tobacco callus but a low ratio of same induced shoot formation. The hypothesis of organogenesis was advanced by Torrey (1966) who opined that organogenesis in callus begins with development of a group of meristematic cells which initiate a primordium. As an outcome of this approach, several hundreds plant species have been reported to form shoot and/or root *in vitro* (Table 1.8).

Table 1.8. Selected examples of regeneration arising from different explants/cultures

(1) Root culture: *Dalbergia sissoo, Vigna aconitifolia*

(2) Shoot tip and axillary bud culture: *Tectona grandis, Morus indica, Piper nigrum, Atropa belladona*

(3) Leaf culture: *Oryza sativa, Rauwolfia serpentina, Azadirachta indica, Hordeum vulgare.*

(4) Embryo culture: *Hordeum vulgare, Allium cepa, Eucalyptus* sp.

(5) Stem culture: *Rose hybrida, Tamarindus indica, Prosopis cineraria*

(6) Inflorescence culture: *Triticale, Musa sp., Zea mays*

(7) Seed culture: *Capsicum* spp., *Cajanus cajan, Vigna mungo*

To achieve organogenesis, optimal conditions of medium constituents are determined empirically. It is apparent that the process of differentiation begins with the changes in individual cells in response to external stimuli. The cells determined to divide and produce callus are changed to produce an organized structure by cellular differentiation. The morphogenetic signals induce the conditions, which allow competent cells to undergo their internally controlled program of differentiation. Such a program stems from selective gene action with the subsequent cellular processes of DNA replication and translation. This results in biochemical and biophysical changes in the target cells, including a change in the metabolic activity. This is followed by cytological and histological changes leading to visible appearance of organ formation. There are sufficient evidences that the process of differentiation begins at single cell. After that centres of meristematic activity occurs surrounding the cell from which the process starts. The organs are developed from these multicellular meristematic centres.

Somatic embryogenesis

It leads to production of bipolar structure containing a root-shoot axis, with a closed independent vascular system. Somatic embryogenesis can be produced directly from the explant (direct embryogenesis) or can be produced via callus formation from explant (indirect embryogenesis). Embryoids have also been artificailly induced in cultured plant tissues, besides zygote. Somatic embryogenesis was first induced in suspension culture (Stewart *et al.*, 1958) and callus culture (Reinert, 1959) of carrot. Somatic embryogenesis may be initiated in two ways (Sharp *et al.*, 1980; Evans *et al.*, 1981): (i) by inducing embryogenic cells within the pre-formed callus, and (ii) directly from pre-embryogenic determined cells (without callus), which are ready to differentiate into embryoids.

The nutritional media of different composition are needed to obtain embryoids. The first medium contains auxin to initiate embryogenic cells. The second medium lacks auxin or reduced level of auxin is required for subsequent development of embryonic cells into embryoids and plantlets. In both the cases a reduced amount of nitrogen is required (Ammirato, 1983). Increased osmotic stress (by using 2 - 6% carbohydrate) in the medium often enhances somatic embryogenesis.

Two theories about induction of somatic embryogenesis have been proposed. According to predetermination theory for somatic embryogenesis, cells undergoing embryo initiation are embryogenic to start with and that the *in vitro* culture conditions simply cause embryogenesis to occur. Further Sharp *et al.* (1980) and Evans *et al.* (1981) suggested that explanted tissues contain cells that are already determined for embryogenic development, namely pre-embryogenic determined cells (PEDC's) and those which require redetermination through a period in culture i.e. induced embryogenic determined cells (IEDC's).

The process occurs naturally in a wide range of species from both reproductive and somatic tissues. Somatic embryos can be found on callus in cell suspensions and protoplast cultures, or directly from cells of organized structures (such as stem segments or zygotic embryo). Around 200 species from angiosperms and gymnosperms have been reported to produce somatic embryos in culture. The first plantlet formation *in vitro* was reported as early as 1940s; Ernest Ball (1946) reported it in *Tropaeolum* and *Lupinus*. There are sevaral plants in which embryogenesis has been induced *in vitro*, some of them are: *Atropa belladona, Brassica oleracea, Coffea arabica, Nicotiana tabacum, Saccharum officinarum*.

There are several advantages of plantlet regeneration through organogenesis or embryogenesis. These include the efficiency of process (the formation of plantlet in fewer steps, simultaneously with reduction in labour, time and cost), the capacity to produce higher number of plantlets, and the morphological and cytological uniformity of the plantlets.

SUGGESTED READINGS

Ammirato, P. V. (1983). *In: Handbook of Plant Cell Culture. Vol 1. Techniques for propagation and breeding.* MacMillan Publ. Co., New York (ed. Evans *et al.*), pp 82-123.

Ball, E. (1946). *Am. J. Bot*, **33**, 301-318.

Bhojwani, S.S. & Johri, B.M. (1970). *Z. Pflanzenphysiol.,* **63**, 269-275.

Bourgin, J.P. & Nitsch, J.P. (1967). *Ann. Physiol. Veg.,* **9**, 377-382.

Carlson, P.S., Smith, H.H. & Dearing, R.D. (1972). *Proc. Nat. Acad. Sci.,* USA, **69**, 2292-2294.

Cocking, E.C. (1960). *Nature,* **187**, 927-929.

Eriksson, T. (1965). *Physiol. Plant.,* **18**, 976-993.

Eriksson, T. & Jonassen (1969). *Planta,* **89**, 85-89.

Evans, D.A. *et al.* (1981). *In: Plant tissue culture: Methods and applications in agriculture.* Acad. Press, New York (ed. Thorpe), pp 45-113.

Gamborg, O.L., Miller, R.A. & Ojima, K. (1968). *Exp.Cell Res.,* **50**, 151-158.

Gautheret, R.J. (1959). *La culture des tissues vegetaux. Masson Cie,* Paris, pp 1-863.

Guha, S. & Johri, B.M. (1966). *Phytomorphology,* **16**, 353-364.

Guha, S. & Maheshwari, S.C. (1964). *Nature,* **204,** 497.

Guha, S. & Maheshwari, S.C. (1966). *Nature,* **212**, 97-98.

Haberlandt, G. (1902). Kulturversuche nut isollierten pflanzenzellen. Sitzungsber. Akad. Wiss. Wien., Math. -Naturwiss, Kl., Abt. 1, pp 69-92.

Johri, B.M. & Bajaj, Y.P.S. (1965). *Phytomorphology,* **15**, 292-300.

Johri, B.M. & Bhojwani, S.S. (1965). *Nature,* **208**, 1345-1347.

Krens, F. A., Molendijk, L., Wullems, G.J. & Schilperoort, R.A. (1982). *Nature,* **296**, 72-74.

Laibach, F. (1925). *Z. Bot.,* **17**, 417-459.

Laibach, F. (1929). *J. Herad.,* **20**, 201-208.

Larebeke *et al.* (1974). *Nature,* **252**, 169-170.

Larkin, P.J. & Scowcroft, W.R. (1981). *Theor. Appl. Genet.,* **60**, 197-214.

Maheswari & Rangaswamy (1958). *Ind. J. Hortic.,* **15**, 275-282.

Melchers, G. & Labib, G. (1974). *Mol. Gen. Genet.,* **135**, 277-294.

Melchers, G., Sacristan, M.D. & Holder, A.A. (1978). *Carlsberg. Res. Commun.*, **43**, 203-218.

Miller, C.O., Skoog, F., Okumura, F.S., Von Saltza, M.H. & Strong, F.M. (1955). *J. Am. Chem. Soc.*, **77**, 2662-2663.

Mitra, G.C. & Chaturvedi, H.C. (1972). *Bull. Torrey Bot. Club*, **99**, 184-189.

Morel, G. & Martin, C. (1952). *C. R. Acad. Sci.*, Paris, **235**, 1324-1325.

Muir, W.H. (1953). *Cultural conditions favouring the isolation and growth of single cells from higher plants in vitro.* Ph.D. thesis, University of Wisconsin, USA.

Muir, W.H. *et al.* (1958). *Am. J. Bot.*, **45**, 589-597.

Murashige, T. & Skoog, F. (1962). *Physiol. Plant.*, **15**, 473-497.

Nitsch, C. (1974). *C.R. Acad. Sci.*, Paris, **278**, 1031-1034.

Nobecourt, P. (1939). *C. R. Seances Soc. Biol. Ses. Fil.*, **130**, 1270-1271.

Paszkowski *et al.* (1984). *EMBO J.*, **3**, 2717-2722.

Pierik, R. L. M. *et al.* (1973). *Scientia Hort.*, **1**, 117-119.

Power, J.B., Cummins, S.E. & Cocking, E.C. (1970). *Nature*, **225**, 1016-1018.

Rechinger, C. (1893). *Abh. Zool. Ges. Wien.*, **43**, 310-334.

Reinert, J. (1959). *Planta*, **58**, 318-333.

Seibert, M. (1976). *Science*, **191**, 1178-1179.

Sharp, W.R., Sondahl, M.R., Caldas, L.S. & Maraffa, S.B. (1980). *Hort. Rev.*, **2**, 268-310.

Shivanna, K. R. (1965). *Phytomorphology*, **15**, 183-185.

Skoog, F. (1944). *Am. J. Bot.*, **31**, 19-24.

Skoog, F. & Miller, C.O. (1957). *Symp. Soc. Exp. Biol.*, **11**, 118-131.

Stewart, F.C., Mapes, M.O. & Mears, K. (1958). *Am. J. Bot.*, **45**, 705-708.

Takebe, I., Labib, G. & Melchers, G. (1971). *Naturwissenschaften*, **58**, 318-320.

Torrey, J.G. (1966). *Adv. Morphogenesis*, **5**, 39-91.

Van Overbeek, J., Conklin, M.E. & Blakslee, A.F. (1941). *Science,* **94**, 350-351.

Vasil, V. (1984). *Cell culture and somatic cell genetics. Vol 1. Laboratory procedures and their applications.* Orlando, Florida, pp 1-825.

Vasil, V. & Hilderbrandt, A.C. (1965). *Science,* **150**, 889-890.

Vochting, H. (1878). *Uber Organbildung im Pflanzenreich.* Max Cohen Publ., Bonn.

White, P.R. (1934). *Plant Physiol.,* **9**, 585-600.

White, P.R. (1937). *Plant Physiol.,* **12**, 803-811.

White, P.R. (1939). *Am. J. Bot.,* **26**, 59-64.

White, P.R. (1953). *Am. J Bot.,* **40**, 517-524.

QUESTIONS

1. Describe various methods of sterilization employed in plant tissue culture work.
2. How Murashige and Skoog's medium is prepared? Write precautionary measures to be taken while praparing the medium.
3. Give a brief account of research work in plant tissue culture done by Indian scientists.
4. Write short notes on the followings:

   (i) Totipotency (ii) Organogenesis (iii) Growth hormones

CHAPTER - 2

SOMATIC EMBRYOGENESIS

When embryos regenerate from somatic cells or tissues (haploid, diploid etc), it is termed as somatic embryogenesis. Embryos can be obtained either directly from cultured explants (the organized structures- like leaf, hypocotyl, stem and other plant parts) and anthers (or pollen) or indirectly from callus and isolated single cells in culture. This is different from zygotic embryogenesis which results from fertilization of an egg cell. The process of embryogenesis involves various stages of differentiation and development such as proembryo, globular, heart shape and torpedo embryo. This phenomenon is most common in Ranunculaceae, Solanaceae, Rutaceae, Umbellifere and Gramineae (Tisserat *et al.*, 1979).

Stewart *et al.* (1958) and Reneirt (1959) reported somatic embryo formation in carrot cell suspension cultures. These somatic embryos were similar to zygotic embryos in development and structure (Fig. 2.1). Although the origin of somatic embryos produced in carrot cell suspension cultures was uncertain, these somatic embryos morphologically developed through the globular, heart and torpedo stages. The development of cotyledons is used to distinguish between mature cotyledonary stage and heart/torpedo shaped embryo. At torpedo stage, cell differentiation occurs establishing root and shoot meristem.

Massive efforts have been put to define the medium conditions, which permit embryogenesis, and characterize the morphological and developmental events leading to embryo formation. Embryogenesis is a two step process. The first step is the induction of embryogenesis while the second step is the development of embryo, ultimately leading to germination. The requirements for embryo induction and embryo development are different, and thus separate media are used for each step. Ammirato (1986) described four stages viz. induction, early growth, embryo maturation and germination or conversion. These stages not only differ in their morphological structure but also in their physico-chemical requirements.

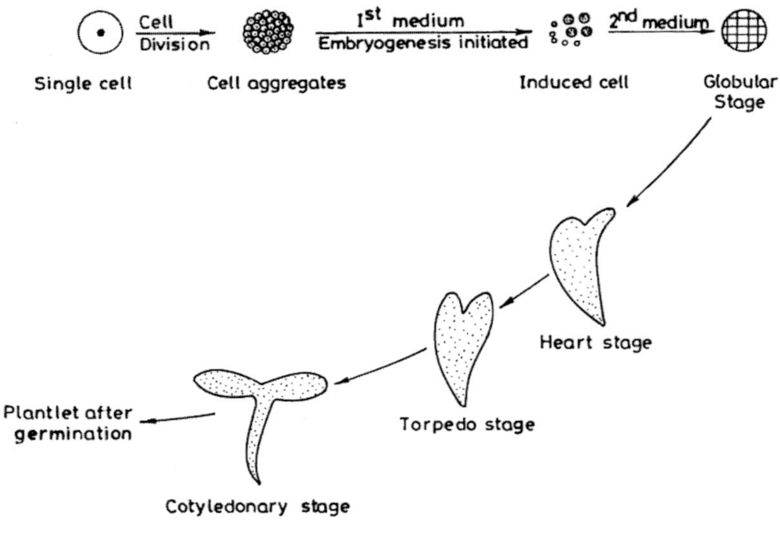

Fig. 2.1. Events of somatic embryogenesis in cultures

Induction

With regard to place of origin, embryogenesis is of two types: direct embryogenesis and indirect embryogenesis (Evans *et al.*, 1981; Sharp *et al.*, 1982; Kohlenbach, 1985; Ammirato, 1986). Juvenile explants like hypocotyls, cotyledons and young immature zygotic embryos are best materials to initiate embryogenic cultures. When embryogenesis occurs directly on the explants without production of callus, it is known as **direct embryogenesis**. When explants produce callus and the callus forms embryos then it is called as **indirect embryogenesis**. An exogenously supplied auxin is needed in appropriate concentration for the induction of somatic embryogenesis from callus or explant. In certain cases (like carrot), exogenous supply of auxin may not be required for embryogenesis to occur. Sharp *et al.* (1982) proposed that in case of direct embryogenesis cells of explanted tissues are already determined for embryogenic development, and these are termed as **pre-embryogenic determined cells (PEDC's)**. The starting material is completely rejuvenated. eg. nucellus of citrus, epidermal cells of hypocotyls (*Ranunculus sp., Brassica napus*). In case of indirect embryogenesis cells require redetermination through a period in culture and this is termed as **induced embryogenic determined cells (IEDC's)**. It has been suggested that this phenomenon is determined by epigenetic factors. Here, differentiated cells must firstly be de-differentiated and then redetermined as embryogenic cells after cell division. Complete rejuvenation must take place. e.g. secondary phloem of carrot, leaf explants of *Petunia, Asparagus, Coffea.*

Embryogenesis occurs from a single cell or from a group of cells. The earliest cell divisions in embryogenetically determined cells follow various patterns, but finally produce embryos of similar shape. Embryogenic cells are small, isodiametric in shape, occupied with dense cytoplasm and have a peculier nucleus. The non-embryogenic cells are relatively large, vacuolated and lack dense cytoplasm. Embryogenic cells from an auxin-containing medium when transferred into a medium having low auxin concentration or without auxin, develop proembryos. Usually media containing 2,4-D, 2,4,5-trichlorophenoxy acetic acid and picloram are used as embryo induction medium. Several species of monocotyledons and dicotyledons have been regenerated through somatic embryo formation by this method. Other auxins like IAA or IBA in combination with a cytokinin have also been found equally good for embryo induction in many dicotyledonous species.

When somatic embryos are transferred on induction medium they give rise to secondary somatic embryos. This method of obtaining embryos recurrently is termed as **repetitive or cyclic embryogenesis**. This method is useful for continuously obtaining embryos in large number e.g. in *Atropa belladonna*, carrot, *Ranunculus, Pennisetum purpurium* and *Pannicum maxicana.*

When young cotyledonary embryos of white clover (*Trifolium repens*) are explanted in presence of cytokinin BAP (6-benzylamino purine), the hypocotyl responds with the formation of somatic embryo. When 2,4-D is used instead of BA, cotyledons form somatic embryos. Likewise, pea shows a positive response from hypocotyl cells in the presence of cytokinin, but it is the cotyledon cells that respond in the presence of auxin. Genotype, tissue type, and developmental stage may influence the comparative ability in responding to auxin or cytokinin.

Development

During somatic embryogenesis in cell suspension cultures embryos of different sizes are produced. Embryos of uniform size are selected. This can be achieved by sieving or fractionation of suspension with appropriate sieve size. Such cultures may be fully synchronized for their growth. The development and maturation of somatic embryos is similar to that of zygotic embryos. Cell differentiation is most evident in the formation of vascular tissues specially visible in the hypocotyl and cotyledon. Differentiated cells should be de-differentiated and then divide, leading to formation of non-organized mass of vacuolated parenchyma cells. This is tranformed into cytoplasm-rich cells that become embryogenic in presence of auxin. The embryoids are formed from embryogenic cells which are small in size, have dense cytoplasmic contents, large nuclei, starch grains and small vacuoles.

Factors affecting somatic Embryogenesis

The composition of culture medium is critical, specially the levels of sucrose and nitrogen:

(1) Nitrate nitrogen alone in high amount is required for induction of somatic embryogenesis. Reduced nitrogen in the embryo development medium supports embryo development.

(2) Favourable effect of certain amino acids like proline and glutamine has been established.

(3) Increasing the osmotic concentration by using high sucrose levels or by addition of mannitol or sorbitol, has shown to affect the embryo development.

(4) Auxins, specially the 2,4-D, appear to be required for embryo induction as in case of cereal crops. Activated charcoal when added to the medium removes excess auxin from the somatic embryos (Buchheim *et al.* 1989).

(5) Cytokinins, with the exception of zeatin, suppress embryogenesis.

(6) Gibberellins and ethylene usually inhibit embryogenesis.

(7) Carrot cells produce somatic embryos below 16% dissolved oxygen while above 16% oxygen levels, roots are produced, in liquid cultures.

(8) The chances of embryogenesis are higher with a callus that has originated from a juvenile plant.

(9) Light promotes embryogenesis, though it can take place at low irradiance or even in darkness in some species.

(10) High temperature favours somatic embryogenesis.

(11) A saccharose concentration of about 2 - 3% and coconut milk promote embryogenesis.

The demonstration of somatic embryogenesis in synchronous cell culture has made this phenonmenon as a model to study organ differentiation and regulation. For instance,

(1) Callus specific and embryo specific proteins are identified. It is observed that embryo specific genes are expressed well before the morphologically visible differentiation.

(2) Rapid increases in the rates of protein and RNA synthesis after transfer of embryogenically competent carrot cells to an auxin-free medium have been shown.

(3) Polyamine synthesis and its role in embryogenesis suggest embryogenic cultures have higher level of polyamines, inhibition of polyamines biosynthesis could suppress embryogenesis, whereas the addition of the polyamine spermidine could reverse this inhibition, and levels of enzymes for polyamine synthesis are higher under embryogenic conditions.

Advantages of Somatic embryogenesis

Somatic embryo is very useful for micropropagation of plant species. A large number of herbaceous dicots and monocots have been regenerated through somatic embryogenesis (Table 2. 1).

(1) Rapid multiplication through production of somatic embryogenesis in cell cultures, and use of bioreactors for scale-up technology.

(2) Somatic embryos grown individually make the system easy to manipulate (to sub-culture).

(3) Provides an important resource for the analysis of molecular and biochemical events that occur during induction and maturation of embryo.

(4) Presence of both root and shoot avoids the rooting step required in organogenesis.

(5) It shortens the breeding cycle of deciduous trees and increases the germination of hybrid embryos.

Table 2.1. Some examples of somatic embryogenesis in tissue culture

(A) **Dicotyledons**: *Arachis hypogea, Brassica napus, Citrus* spp., *Coffea arabica, Glycine max, Mangifera indica, Santalum album, Daucus carota, Hevea* sp., *Ipomoea batata* etc.
(B) **Monocotyledons**: *Allium sativum, Musa* sp., *Cacos nucifera, Elaeis guineensis, Avena sativa, Oryza sativa, Secale cereale, Zea mays* etc.

Disadvantages of Somatic embryogenesis

(1) High risk of mutations involved.

(2) Repeated sub-culturing may weaken regenerative capacity.

(3) Induction of embryogenesis sometimes is difficult or impossible with many plant species.

Potential Applications of Somatic embryogenesis

(1) Mass propagation: Somatic embryos have powerful advantages for mass propagation in comparison to both conventional clonal propagation methods (rooted cutting, grafting) and other *in vitro* regenerated systems (e.g. in propagation). Unlimited number of embryos can be produced from a single explant.

(2) Raising somaclonal variants in tree species is possible.

(3) Source of regenerable protoplast system: Embryogenic callus and suspension cultures, as well as somatic embryos themselves have been employed as a source of protoplasts for a range of species. e.g. in three groups of plant species, viz., graminaceous species, citrus species, and forest trees (specially conifers).

(4) Embryo culture and gene transfer: Regeneration in several species, specially trees and large-seeded legumes, is limited to regeneration via direct somatic embryogenesis from immature zygotic embryos. The embryos form directly on the original explant tissue. If callus is present, it grows concomitantly with the somatic embryos. The embryos do not originate from it, thereby by-passing any chance that a callus phase provides to sort transferred cells from non-transferring cells. So callus phase can be by-passed through repetitive somatic embryogenesis. McGranahan *et al.* (1990) exploited repetitive embryogenesis for agrobacterium-transformed walnut cells to obtain multiple plants of somatic embyros by-passing the callus phase.

(5) Clonal propagation through somatic embryogenesis is reported in about 60 species of woody trees.

(6) Somatic embryogenesis provides potential plantlets in the form of somatic seeds. The somatic embryo can be used for the production of synthetic seeds for direct sowing in the field. This gives impetus to increased agricultural production.

(7) Somatic embryo provides organized culture system to produce organ specific or differentiation related compounds in higher amounts as in case of *Digitalis* and *Theobroma cacao*. Repetitive somatic embryogenesis in borage ensures a good availability or supply of ϒ-linolenic acid, which is used in the teatment of atopic eczema.

SUGGESTED READINGS

Ammirato, P.V. (1986). *Hort. Sci.*, **21**, 843.

Evans, D. A. *et al.* (1981). *In*: *Plant tissue culture. Methods and applications in agriculture* (ed. Thorpe). Acad. Press, New York, pp 45-113.

Kohlenbach, H.W. (1985). *In: In vitro techniques. Propagation and long term storage* (ed. Schafer-Menuhr), Netherlands, pp 101-109.

McGranahan, G. H., Leslie, C. A., Uratsu, S. C. & Dandekar, A. M. (1990). *Plant Cell Rep.*, **8**, 512-516.

Reinert, J. (1959). *Planta*, **58**, 318-333.

Sharp, W. R., Evans, D. A. & Sondahl, M. R. (1982). *In: Plant tissue culture* (ed. Fujiwara, A.). Proc. 5th Intl. Congr. Plant Tiss Cell Cult, Japan, pp 759-762.

Stewart, F. C., Mapes, M. O. & Mears, K. (1958). *Ann Bot.*, **16**, 57-77.

Tisserat *et al.* (1979). *Hort. Rev.*, **1**, 1-78.

QUESTIONS

1. What is somatic embryogenesis? How induction and development aspects of somatic embryogenesis influence the process.
2. Discuss the advantages and applications associated with somatic embryogenesis.
3. Write short notes on the followings:

   (i) Role of growth hormones in somatic embryogenesis

   (ii) Factors affecting somatic embryogenesis

CHAPTER - 3

MICROPROPAGATION

The classical methods of *in vivo* vegetative propagation often fall short of that required (too slow and difficult, too expensive) or are absolutely impossible. *In vitro* propagation technology can overcome all such problems. It has now become possible to clone species by *in vitro* culture techniques that are impossible to clone *in vivo*. One of the aspects of plant tissue culture is the ability to produce a large number of identical individuals via *in vitro* cloning, and it is popularly called as 'micropropagation'. Micropropagation in tissue culture is achieved by three ways:

(i) *Axillary budding*: Development of shoots from pre-existing meristems on nodal region ensures genetic stability of the regenerants and is termed as axillary budding. About 90% of the current production of the micropropagated plants is through axillary budding and apical shoots. Propagation of long (life) cycle plants such as forest and fruit trees is multiplied mainly through this method. This approach has been used in case of strawberry, potato, asparagus, pineapple, banana, etc. Apical shoots (1-5 mm) are normally cultured on culture media having mixtures of auxin (0.01 - 0.1 mg l^{-1}) and cytokinin (0.05 - 0.5 mg l^{-1}).

(ii) *Adventitious shoots*: These are stem, leaf structure, bulbs, rhizomes, tubers, etc. The meristematic regions regenerate multiple shoots on a suitable culture medium.

iii) *Somatic embryogeny*: It is the most efficient method. It does not ensure genetic stability. However, somatic embryo in cell cultures is very useful for developing scale-up technology using bio-reactor and in the production of synthetic seeds through encapsulation. Somatic embryogeny has been done in about 200 species of different groups of plants.

Procedures of Micropropagation

The two main approaches to micropropagation are:

(i) multiplication of shoot meristem and (ii) serial sub-culture of stem nodes.

In vitro clonal propagation requires many stages:

Stage I - III are done under *in vitro* conditions, whereas Stage IV is achieved in a greenhouse environment. An additional Stage 0 has been suggested for various micropropagating systems.

Stage 0: Selection and maintenance of stock plants for culture initiation including correct pre-treatment of starting materials.

Stage I: Explant isolation, sterilization and establishment on a suitable culture medium.

Stage II: Multiplication of shoots using an appropriate nutrient medium (i.e., propagation phase).

Stage III: Rooting of *in vitro* regenerated shoots. i.e., preparation of shoots and plants for transfer into soil. This involves stopping the axillary shoot formation and initiating shoot elongation, subsequently root formation is to be induced.

Stage IV: Transfer of plantlets to sterilized soil for hardening under greenhouse environment.

The detailed methodolgy/procedure for micropropagation is given below:

(1) Shoot tip/nodal stem portion is the ideal starting explant for micropropagation.

(2) The explant is surface sterilized, then trimmed, and the remaining material is then placed on a variety of nutrient media, under aseptic conditions.

(3) This is the phase for acclimatization of the inoculated tissue to the *in vitro* conditions that requires about 3 or 4 months and varies from species to species.

(4) Under suitable laboratory conditions, the shoot tip grows giving rise to a small unrooted shoot in culture.

(5) This *in vitro* grown shoot, after being on the initiation medium for a suitable period, is sub-cultured onto a multiplication medium. Multiplication medium has higher levels of cytokinins.

Sub-culturing can be essential for various reasons: (i) the nutrient medium is exhausted (ii) the nutrient medium dries out, resulting in higher salt and sugar concentrations (iii) brown/black colouring appears in the agar due to secretion of toxic substances by plant tissues. (iv) the medium becomes liquid as a result of lowering of pH by the plant. Usually the period between sub-cultures is 3 weeks; by this period number of shoot tips increases many-folds owing to breaking of nodal axillary buds. This again provides materials for further sub-culturing.

(6) Rooting of these cultured (sub-cultured) shoots is possible by transferring to auxin-rich media that induce formation of *in vitro* roots. Rooting is followed by weaning to a normal humidity level. Rooting requires low light, and hence plants must not be exposed for the first time to full sunlight. Such hardening/ acclimatization of *in vitro* grown plants, before transferring in pots in the field, can also be achieved by keeping them in plant growth cabinets where the environmental conditions can be manipulated from lower to higher degree of exposure to light, temperature and humidity.

(7) All these procedures help develop small photosynthesizing plants.

Hundreds of plants have been micropropagated (Table 3.1). Use of a phenolic compound called phloroglucinol (PG) found in xylem sap of apple has been reported to be of significance for shoot multiplication and rooting in a number of rosaceous fruit cultivars (Jones, 1983). Fig. 3.1 shows various steps involved in micropropagation of a plant.

Table 3.1. Examples of plants that are micropropagated on mass scale

Ornamentals	Woody species	Other species
Bromeliads	*Araucaria*	*Actidinia*
Chrysanthemum	*Betula*	*Allium*
Cymbidium	*Coffee*	*Asparagus*
Dianthus	*Eucalyptus*	*Cardamom*
Ferns	*Populus*	*Festuca*
Gladiolus	*Pyrus*	*Glycine*
Orchids	Rose	*Phaseolus*
Rhododendron	*Sequeoia*	*Solanum*
Saxifraga	*Tectona*	*Trifolium*
Pelargonium	*Vitis*	*Musa*

Factors affecting micropropagation

(1) Plant health and physical status of plant explant: Explants are to be used from a good healthy plant; plant material should be tested for bacterial and virus indexing prior to micropropagation. Senescing tissues are avoided for culturing. Plant materials isolated from recently produced parts of a plant are more regenerative than those from older regions. Physiological state of plant has an important bearing on *in vitro* cell division and regeneration.

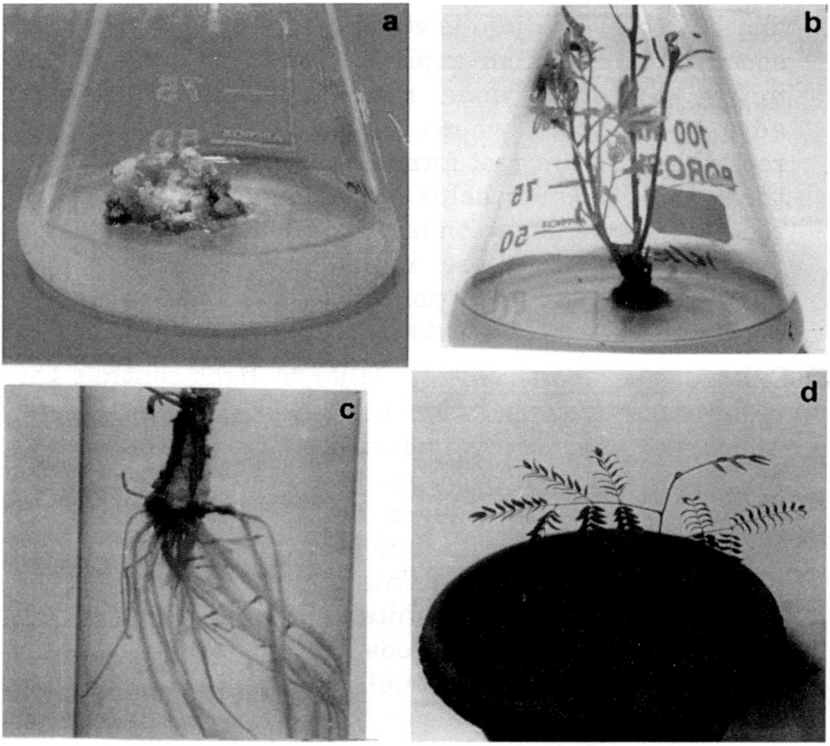

Fig. 3.1. Different stages of Micropropagation of Leucaena leucocephala ***a.*** *callus formation,* ***b.*** *multiple shoot regeneration,* ***c.*** *rooting stage;* ***d.*** *field transfer of planted to pot. Courtesy: Prof. U.N. Dwivedi, Deptt. of Biochemistry, Lucknow University, Lucknow.*

(2) Culture nutrient medium: A medium which does not favour callus formation is considered to be suitable as the genetic instability of callus will lead to a high degree of genetically aberrant plants (Dodds & Roberts, 1985). Whereas standard tissue culture media are more suitable for achieving stages I and II of micropropagation, only stage III needs certain modifications. The mineral components as well as two types of growth regulators (auxin and cytokinins) influence the success of micropropagation of a culture. The amount of NH_4^+

and K^+ in the medium governs the number and size of shoots produced from explants. Auxin is preferred in nutrient medium when the intension is to obtain adventitious roots, whereas gibberellins and ABA are reported to inhibit root formation. Activated charcoal in the culture medium helps in induction of adventitious roots, removes the secondary metabolites from culture media which otherwise inhibit the explant. Media having a low concentration of salts favour rooting of shoots micropropagated at stage II.

(3) Culture environment: The quality of light controls the organogenic differentiation and shoot growth *in vitro*. Morphogenesis during micropropagation is influenced by red/far red light exposure. The light illumination of 16 h day and 8 h dark gives the best result in terms of multiplication and proliferation of shoots, though there are some exceptions to it. This photoperiod provides the average requirements of white light for shoot induction, whereas shorter photoperiods are needed to stimulate rooting. Temperature is usually maintained constant at 25-27°C. Sometimes, depending on species a lower temperature (e.g. 18°C for bulbous species), or higher temperature (28-29°C for tropical species) is selected. Very special temperature requirements are also followed, for organogenesis (Pierik, 1967). Alternating temperature conditions may also be needed, as in case of *Helianthus* sp., where rooting is favoured by a day temperature of 26°C and night temperature of 15°C.

Oxygen supply in culture tubes is maintained by use of metal caps, apolar inoculation and use of liquid media. Frequent shaking of culture media on a shaker, and orchid wheels - fitted with flasks having liquid media help facilitate aeration. Each morphogenic response needs an optimal oxygen concentration for a particular organ and species (Norton, 1988). Carbondioxide in excess (up to 20%) may induce dwarfing or albinism in the shoots of *Cordyline terminalis*, *Ficus* sp. (De Proft *et al.*, 1985). Ethylene is known to affect different morphogenic responses in plant tissue cultures, like callogenesis, embryogenesis and rhizogenesis. It can affect potato tuberization *in vitro*, and shoot elongation in roses (Hussey & Stacey, 1984, Kevers *et al.*, 1992). However, ethylene can bring about positive effect as well. For instance, it can suppress apical dominance, increased axillary shoot production, increased bulb primordia development in regenerated shoots of tulips, etc (De Proft *et al.*, 1985, Taeb & Alderson, 1990).

(4) Genotypes: Dicots can regenerate more effectively than monocots, and gymnosperms have little regenerative capacity. Dicot plants from Solanaceae, Crassulaceae, Cruciferae regenerate much easily.

(5) Position of explant within the plant: Higher the shoot isolated from a tree, the lower is probability of adventitious root formation.

(6) Method of inoculation: Explants are placed on nutrient medium in various ways - polar (physiological base in the medium) or apolar (physiological base out of medium and top in medium). The latter way of inoculation facilitates better regeneration of roots and shoots (Pierik & Steegmans, 1975).

(7) Contamination by pathogens: Bacteria (*Pseudomonas sp., Bacillus sp. and Erwinia sp.*) are major contam-inants during micropropagation. Addition of antibiotics (like ampicillin to kill bacteria or fungicides like Bevistin to kill fungal contaminants) may control the infection by these microbial contaminants. These anti-biotics are added to the medium by filter sterilization. However, the addition of antibiotics often leads to phytotoxic effect.

(8) Browning of the culture media: It is the most frequent problem associated with micropropagation of woody perennials. This is caused by accumulation of inhibiting

compounds (phenolics) in growth medium, which turn the medium dark brown. These substances are toxic to tissues and inhibit their growth. This can be prevented by:

(a) adding activated charcoal (0.2-3.0%, w/v) to the medium. Activated charcoal is obtained after wood is carbonised at high temperature in presence of steam. It has minute pores with large inner surface area on which gases and solid compounds can be adsorbed and thus removed.

(b) adding polyvinylpyrrolidone(PVP) to the medium at a concentration of 250 - 1000 mg l^{-1} (Johansson, 1983).

(c) adding anti-oxidants like citric acid, ascorbic acid, L-cystine.

(d) adding glutamine, arginine and asparagine.

(e) frequent sub-culturing onto a fresh medium at short intervals.

(f) soaking explants in water before placing in culture - this helps reduce exudation of toxic substances from explants into the culture medium.

Applications of Micropropagation

(1) Micropropagation of Tree species

Due to problems associated with the conventional methods of propagation (like non-availability of seeds, infestation of seeds by pests and pathogens, and recalcitrant seeds), quality planting materials in tree species are not available. The tissue culture techniques, however, offer means for mass multiplication as well as conservation of fruit and forest trees and other elite and rare trees. Micropropagation *in vitro* has been done successfully in many Himalayan plants. For instance, (i) proliferation of shoots from terminal and axillary buds as obtained in *Camellia sinensis* (Palni *et al.*, 1991) and citrus spp. (Bhat *et al.*, 1992). (ii) axillary shoots raised from nodal segments of

Bauhinia variegata (Mathur & Mukunthakumar, 1992), *Dendrocalamus brandisii* (Nadgauda *et al.*, 1990) and *Dendrocalamus strictus* (Chaturvedi & Sharma, 1988). Use of BAP (10 μM) and IBA (1 μM) to cultured nodal segments enhances shoot multiplication at rapid speed (about 12 - 27 folds) within two months (Bag *et al.*, 1997).

(2) Micropropagation of medicinal plants

The use of micropropagation techniques for medicinal plants gives many benefits to the breeders due to various reasons: (a) it leads to mass multiplication of those plants whose seeds have low regeneration or which fail to produce seeds in a given climate (b) it makes available plants throughout the year (c) it enables the availability of uniform clones from highly heterozygous plants, and (d) it helps conserve the plant germplasms which are on the verge of being endangered. Various explants like stem and leaf segments, shoot buds, hypocotyl, root, anthers etc have been tried successfully to regenerate medicinal plants *in vitro* (Table 3.2).

Table 3.2. Micropropagation as employed in some medicinal plants

Plants	Explant used	Medicinal value	Response Observed	Reported by
Artimisia annua	Nodes	Anti-malarial	Plantlets	Gulati *et al*, 1996
Santalum album	Nodes	Essential oil	Callus, plantlets	Lakshmi Sita, 1986
Catharanthus roseus	Stem	Anti-leukaemia	Plantlets	Datta & Srivastava, 1997
Datura innoxia	Anther	Anti-cholinergic	Plantlets	Srivastava *et al.*, 1993
Delonix regia	Leaves	Agglutin-ation	Plantlets	Gupta *et al.*, 1996
Moringa oleifera	Hypocotyl	Anti-viral	Plantlets	Mohan *et al.*, 1995

Panax ginseng	Anther	Tonic	Plantlets	Du *et al.*, 1986
Taraxacum officinale	Root	Hepatic stimulant	Shoots	Bowes, 1970
Rauwolfia serpentina	Plant parts	Sedative, Hypotension	Plantlets	Chaturvedi *et al.*, 1982
Hyoscyamus niger	Seedling	Anti-spasmolytic	Plantlets	Cheng & Raghavan, 1985

(3) Micropropagation of Ornamental plants

Micropropagation of various ornamental plants have been achieved, for instance, *Chrysanthemum, Hedera, Iris, Cymbidium, Vanda, Dendrobium, Agave, Gladiolus, Dianthus*, etc, and the explants used for *in vitro* culture in regeneration of these plants include axillary buds, apical shoots, inflorescence, leaves and roots (Sagawa & Kunisaki, 1982; Robert *et al.*, 1987).

Advantages of Micropropagation

(1) Small surface needed.

(2) Production can be maintained without seasonal variation. As a result of perfect conditioning of nutrient media and physical conditions, enabling precise timing of cutting production, the effect of seasons can be eliminated and year round production can be achieved.

(3) Stock plants can be stored *in vitro*.

(4) Plants difficult to be propagated *in vivo* can be propagated *in vitro*. This is sometimes due to rejuvenation, which is only possible *in vitro*.

(5) No attention is required between sub-cultures (e.g. watering, weeding, etc.).

(6) Virus free plants can be produced.

(7) Propagation *in vitro* is more rapid than *in vivo*.

(8) Growth of *in vitro* propagated plants is often stronger than those cloned *in vivo*.

(9) Micropropagation can produce great savings in fuel costs, greenhouse space, etc.

(10) The propagation *in vitro* on its own root system makes grafting and budding onto root stocks unnecessay - it has importance for species like *Rose, Lilac*, etc.

Disadvantages of Micropropagation

(1) Experience and specialized facilities needed.

(2) Skilled operation required.

(3) If fungal/bacterial contamination occurs during early multiplication many potential propagules may be lost.

(4) Propagation technique can introduce genetic instability.

(5) Initial plantlets are small.

(6) An occasional problem with repeated cycles of *in vitro* shoot is the development of swollen, distorted leaves which become irreversibly translucent and eventually necrotic, a condition that may lead to death of shoot cluster. This is called 'water soaking' by Hussey, and later on described as "Vitrification" by Debergh *et al.* (1981). In literature, vitrification is termed as translucency, glauciness, hyperhydration, water logging. This is particularly common if the plant has excess water available, or if the medium has a low agar concentration. This phenomenon has been described in *Malus* (Anonymus, 1978), in carnation (Ziv & Halevy, 1983) and artichoke (Debergh *et al.,* 1981). Its occurrence and degree is governed by many factors, notable amongst them are:

(a) increasing the concentration of agar and/or sugar lessens the vitrification, as observed in case of carnation and artichoke.

(b) it can be avoided in some cases by better gaseous exchange in culture vessels.

(c) it is associated with high cytokinin levels (Hussey, 1986).

(d) it is induced at low light irradiance and high temperatures.

(e) young plant explants (i.e. softer) are more susceptible to vitrification.

(f) it may be reduced by changing the mixture of macro salts.

(7) Plants produced through this technology may show troublesome effects *in vivo*; repeated production of side shoots and complete reversal to juvenile phase.

(8) Rooting in case of woody plants is often difficult.

(9) Transfer of some plants from test tube (laboratory) to soil is very difficult to achieve.

(10) Regenerative capacity may be lost by repeated culture of callus and cell suspensions.

SUGGESTED READINGS

Bag, N., Palni, L.M.S. & Nandi, S.K. (1997). *Physiol. Mol. Biol. Plants*, **3**, 99-103.

Bhat, S. R., Chitralekha, P. & Chandel, K.P.S. (1992). *Plant Cell Tiss. Org. Cult.*, **29**, 19-25.

Bowes, B. G. (1970). *Protoplasma*, **71**, 197-202.

Chaturvedi, H. C. & Sharma, M. (1988). *Proc. 75th Indan Sci. Cong.*, p. 297.

Chaturvedi, H. C., Sharma, A. K., Sharma, M. & Prasad, R. N. (1982). *In: Plant Tissue Culture* (ed. Fujiwara, A.). Maruzen, Tokyo, pp. 687-688.

Cheng, J. & Raghavan, V. (1985). *Am. J. Bot.*, **72**, 580-587.

Datta, A. & Srivastava, P. S. (1997). *Phytochem.*, **46**, 135-137.

De Proft, M. P., Maene, L. J. & Debergh, P. C. (1985). *Physiol. Plant.*, **65**, 375-379.

Debergh *et al.* (1981). *Physiol. Plant.*, **53**, 181-187.

Dodds, J.H. & Roberts, L.W. (1985). *Experiments in Plant tissue culture*, 2nd ed. Cambridge University Press, Cambridge.

Du, L.G., Dunford, H. & Hou, Y.H. (1986). *In: Proc. 6th Intl. Congr. Plant Tiss. Cell Cult.,* Minneapolis, Abstr. 98.

Gulati, A., Bharel, S., Abidin, M.Z., Jain, S. K. & Srivastava, P. S. (1996). *J. Biochem. Biotech.,* **5**, 31-35.

Gupta, N., Jain, S.K. & Srivastava, P.S. (1996). *Phytomorphology*, **46**, 267-275.

Hussey, G. & Stacey, N. J. (1984). *Ann Bot.*, **53**, 565-578.

Johansson (1983). *Physiol. Plant.*, **59**, 397-403.

Jones (1983). *Biologist*, **30**, 181-188.

Kevers, C., Boyer, N., Courduroux, J. C. & Gasper, T. (1992). *Plant Cell Tissue Organ Cult.*, **28**, 175-181.

Lakshmi Sita, G. (1986). *In: Biotechnology in agriculture and forestry* (ed. Bajaj,Y.P.S.). Springer Verlag, Berlin, pp 363-374.

Mathur, J. & Mukunthakumar, S. (1992). *Plant Cell Tiss. Org. Cult.*, **28**, 119-121.

Mohan, V., Purohit, M. & Srivastava, P.S. (1995). *Phytomorphology*, **45**, 253-261.

Nadgauda, R. S., Parasharami, V. A. & Mascarenhas, A. F. (1990). *Nature*, **344**, 335-336.

Norton, C. R. (1988). *Acta Hortic.*, **227**, 302-304.

Palni, L. M. S., Sood, A., Chand, G., Sharma, M., Rao, D. V. & Jain, N. K. (1991). *In: Proc. Intl. Symp. Tea Sci.,* Shizuoka, Japan, pp 395-399.

Pierik (1967). *Meded. Landbouwhogeschool Wageningen*, **67**, 1-71.

Pierik & Steegmann (1975). *Scientia Hort.*, **3**, 1-20.

Robert, M. L., Herrera, J. L., Contreras, F. & Scorer, K. N. (1987). *Plant Cell Tiss. Org. Cult.*, **8**, 37-48.

Sagawa, Y. & Kunisaki, J. T. (1982). *Jpn. Ass. Plant Tiss. Cult.*, Tokyo, pp 683-684.

Srivastava, P. S., Purohit, M., Pande, D. & Datta, A. (1993). *Phytomorphology*, **43**, 209-216.

Taeb, A. G. & Alderson, P. G. (1990). *Hort.Sci.*, **65**, 199-204.

Ziv & Halevy (1983). *Hort. Sci.*, **18**, 434-436.

QUESTIONS

1. What is micropropagation? Describe various procedural steps by which it is achieved.
2. Write briefly about factors which influence micropropagation.
3. Write short notes on the followings:
   (i) Micropropagation of medicinal plants
   (ii) Advantages of micropropagation
   (iii) Limitations of micropropagation.

CHAPTER - 4

PROTOPLAST CULTURE AND SOMATIC HYBRIDIZATION

Protoplasts are plant cells with a plasma membrane but no cell wall. Separation of plant cells and removal of the cell wall is done experimentally, by either a mechanical or an enzymatic process, giving rise to the production of isolated and naked plant cells, termed as 'isolated protoplasts'. Protoplasts thus can be referred to as 'functional individual cell with plasma membrane as outermost layer and the only barrier between environment and the cell contents'.

Pioneering developments

(1) Klerckel (1892) tried to isolate protoplasts from plasmolysed higher plant cells by mechanically cutting the cells. (2) The impetus to protoplast research was given in 1960's with the enzymatic removal of cell wall by E.C. Cocking at the University of Nottingham, using cell wall degrading enzymes. (3) Takabe *et al.* (1968) with the help of commercially available enzymes isolated protoplast and able to regenerate whole tobacco plant from mesophyll protoplasts. (4) Carlson *et al.* (1972) were successful in obtaining the first interspecific somatic hybrids using protoplast fusion between *Nicotiana glauca* and *N. longsdorffii*. Polyethyleneglycol (PEG) was used as a potent fusogenic agent. With subsequent discovery of fusogens, production of several somatic hybrids including the first

intergeneric hybrid (pomato) between potato and tomato was demonstrated by Melchers *et al.* (1978). (5) Zelcer *et al* (1978) obtained the first interspecific hybrid between protoplasts of normal *Nicotiana sylvisteris* and x-ray irradiated protoplast of male sterile *N. tabacum*. In 1979, stimulation of protoplasts fusion by electrical stimulus and gene transfer through protoplast fusion were shown. During 1962-1975, basic culture media and protocols were established for protoplasts, cell and tissue culture. During 1979-1984, a method was developed for mass selection of hybrid cells using fluorescence-activated cell sorting (Harkins & Galbraith, 1984). (6) Several transgenic plants are produced by using polyethylene glycol mediated fusion (Krens *et al.*, 1982) and DNA uptake by electroporation (Fromm *et al.*, 1985). In 1987, using computer – directed system, selection, culture and manipulation of protoplasts was achieved (Schweiger *et al.*, 1987). The first release of a commercial cultivar obtained from protoplast fusion was in 1990 (Brown & Thorpe, 1995).

Isolation of Protoplasts

Accomplishments in somatic hybridization involve many stages/phases that are required to be followed:

(1) Choice of plant material. Different explants are used for the purpose (Table 4.1).

(2) Sterilization of the explant, in case it has not been grown *in vitro*.

(3) Enzymatic treatment to obtain protoplasts and their purification.

(4) Protoplast fusion followed by selection of somatic hybrids.

(5) Culture of protoplasts, cell wall regeneration and induction of cell division.

(6) Regeneration of plants.

(7) Verification and control of fusion products.

(8) Testing the usefulness in plant breeding.

Table 4.1. Sources of plant tissues from which protoplasts can be isolated

Plant tissues	Young leaves, roots, stems
Flower parts	Petal, reproductive organs
Callus/cells	Friable callus tissues/cells
Suspension cultures	Fast growing cells
Aseptically grown plantlets	Leaves, stems, roots

Protoplasts are isolated by two procedures: mechanical and enzymatic. The former method is no more in use. This involved cutting of plasmolysed cells. Protoplasts using this method were isolated from vacuolated cells of storage tissues. e.g. onion bulbs, radish root, mesocarp of cucumber. The enzymatic method of protoplast isolation is given below:

Isolation of protoplasts from leaves

It involves four steps: (i) Sterilization of leaves (ii) Peeling off the lower epidermis of leaves. (iii) pre-enzyme treatment and incubation in enzyme, and (iv) isolation and cleaning of the protoplasts by filtration and centrifugation.

Isolation of protoplasts from callus and cell cultures

In vitro grown cultures are used directly for the isolation of protoplasts. In fact, actively growing cultures in the exponential phase are good material for the isolation of protoplasts. Normally one gram material is used to obtain protoplast and conditions are optimized empirically by altering enzyme concentration and composition, concentration of osmoticum, duration of incubation and types of tissues.

Protoplasts are obtained by enzymatically treating mesophyll cells, callus, cell suspensions, etc with a mixture of enzymes (cellulose, pectinase and hemicellulose). These enzymes dissolve cellulose, pectin (middle lamella) and hemicellulose, respectively. During the enzyme treatment, the protoplasts obtained need to be stabilized. Therefore, an osmoticum - glucose, mannitol or sorbitol is added which prevents protoplasts from bursting. The enzyme solution (pH having 5.5) is prepared in 10-15% sorbitol/mannitol for membrane stability, and this enzyme treatment should be as short as possible. Other isolation conditions that influence protoplast isolation are temperature, light conditions, speed of shaking, etc; enzymic treatment is done in the dark. Petri plates containing tissue and enzyme mixture are sealed with parafilm and incubated for 4 -12 h on orbitol shaker at 25°C. After incubation, solution is filtered through nylon mesh (50-100mm) which removes debris, transferred into centrifuge tubes, and centrifuged at 100 g. The protoplasts form a pellet, whereas the debris in the supernatant is carefully removed. Fresh sterilized sorbitol solution (no enzyme) is added to tube and centrifuged. By repeating the process 2 to 3 times, protoplasts are cleared (debris removed). Use of 20% sucrose solution causes protoplasts to float and debris settle during centrifugation at 200 g for 1 min. Floating protoplasts are carefully removed using sterilized pipette. Protoplasts are counted by haemocytometer and then diluted to proper strength in the culture medium containing osmoticum.

Purification of Isolated Protoplasts

After protoplasts are obtained they are to be purified. Following steps are involved:

(i) *Sedimentation and Washing*: Protoplast suspension is centrifuged at low speed (50-100 × g/5 min); intact protoplasts form pellet. The supernatant having debris, cell wall, and all organelles, etc. are then pipetted off. The pellet is then resuspended in fresh culture containing mannitol and rewashed.

(ii) *Floatation*: Owing to low density of intact protoplasts compared to cell organelles, many types of gradients have been used that allow the protoplasts to float and for the sedimentation of cell debris. A concentrated solution of mannitol, sorbitol or sucrose (0.3 - 0.6 mM) may be combined with the enzyme-protoplast mixture and centrifuged at low speed. Protoplasts are pipetted off.

Protoplast Viability Test

Isolated protoplasts should be healthy and viable in order to experience sustained divisions and regeneration. This is done in following ways:

(1) Exclusion of Evans Blue dye (2) Presence of cytoplasmic streaming (3) Change in protoplast size induced by changes in the level of osmoticum, and measure of photosynthetic and respiratory activities. These tests indicate active metabolism in protoplasts. Dyes like fluorescein diacetate (FDA), phenosafranin and calcofluor white (CFW) are also popularly used for the purpose. As the FDA accumulates within the plasma membrane, viable protoplasts fluoresce green or white. Protoplast preparation treated with FDA (normally 0.01%) is normally observed within 5 - 15 min, because after this period FDA dissociates from the membrane.

Factors influencing protoplast cultures

(1) Isolated protoplasts require somatic protection in the culture medium until they generate a strong wall. Prolonged culture at high osmotic pressure can result in browning of the cultures and inhibition of callus growth. The osmolarity of the medium is gradually decreased with cell wall formation and cell division.

(2) Nutritional components also affect the process. Some modification can have beneficial effects on protoplasts'

survival and colony formation; C, N, vitamins and plant growth regulators are suitably modified. Concentration of nutrients is crucial. Presence of ammonium in culture medium is toxic to protoplasts of potato, several fruit tree species and some species of compositae. Increased Ca^{2+} concentartion helps membrane integrity (Torres, 1989). Table 4.2 gives modified B5 medium with some of the nutrient's concentration altered.

Table 4.2. Modified B5 medium (Kartha *et al.*, 1974)

Constituents	Concentration	Constituents	Concentration
Sorbitol	4.5% (w/v)	$CaCl_2.2H_2O$	0.0875%
Mannitol	4.5% (w/v)	2,4-D	2.3×10^{-6} M
d-glucose	0.25% (w/v)	BAP	10^{-6} M
D-ribose	0.125% (w/v)	NAA	10^{-6} M

(3) Growth regulators: Both auxin and cytokinin are needed, exception being protoplast system of *Citrus* and *Petunia* tumor protoplasts. It is recommended to change growth regulator supply after the first few divisions, as the hormonal requirement of protoplasts and protoplasts-derived cells may be different. Protoplasts from actively growing cell cultures require high auxin/kinetin ratio for inducing divisions, whereas those derived from highly differentiated cells (like leaf tissue) need a high kinetin/auxin ratio for regeneration.

(4) Several environmental factors also affect protoplast culture. For instance, high light intensities sometimes inhibit protoplasts division, therefore, protoplasts are cultured in low light or darkness; the suitable temperature for culturing protoplasts is 24 - 28°C. A short heat treatment at 45°C for 5 min improved the protoplats division in wheat, rice and *Pennisetum*; low voltage electric pulse treatment is known to stimulate the division of protoplasts, as observed in *Trifolium;* pH 5.5 to 5.8 is found suitable; after production of

microcalli, sub-culture on to another medium induces regeneration.

There are some special culture methods in use which enhance the efficacy of the protoplast culture:

(i) Immobilization of plant protoplasts: Immobilization of protoplasts improves survival and cell division in several difficult plant materials. For instance, in case of barley, sunflower and maize, immobilization of protoplasts is done to obtain cell division. The gel matrix provides mechanical support to the fragile cell and inhibits the lipid peroxidation of the membrane.

(ii) Nurse culture and conditioned media: The nutrient media used for protoplast culture may not provide exact nutrition needed by single cells. Nurse tissue culture technique and conditioned medium for the growth of protoplasts can help improve the cell growth of protoplasts, as used in case of wheat, *Festuca, Brassica* protoplast cultures.

Protoplasts fusion and Somatic Hybridization

Incompatibility at various levels, before fertilization and after fertilization restricts recombination between distantly related species. Protoplasts fusion provides a novel approach in combating sexual incompatibility and obtaining somatic hybrids. Fusion of somatic cells (2n) and production of hybrids is known as somatic cell fusion/somatic hybridization. Kuster (1909) proposed the protoplasts fusion as a tool for production of hybrids.

The protoplasts fusion requires the direct contact between membrane surfaces. The contact of protoplasts depends on surface charge. The outer membrane surface is negatively charged and varies from -10 mV to -50 mV and can significantly be reduced by Ca^{2+}. Positively charged protoplasts can be obtained in the presence of poly-L-lysin, synthetic phospholipids, etc. After establishment of contact between two protoplasts, membrane fusion results by structural modifications and can be felicitated by fusogenic agents. As a reult of protoplast fusion, it has been possible to achieve protoplast fusion between dicot and dicot, dicot and monocot, plant and insect cells, etc. A number of plants have been regenerated (Table 4.3).

Fusion Procedure

Protoplasts of both types are mixed in an equal proportion attaining a density of 5×10^4 to 2×10^5 protoplasts per ml. There are several chemical and physical agents, known as fusogen or fusion agent which help in fusion between protoplasts, e.g. sodium nitrate, polyethylene glycol of different molecular weight (PEG 1500, PEG 4000, PEG 6000), dextran sulphate, gelatin, etc. High Ca^{2+} ion concentration and pH (10.5) of the incubation mixture facilitate fusion process. Fusion is stimulated under electric charge as membranes are negatively charged (Zimmermann, 1982).

Isolated protoplasts are suspended in an aggregation mixture (5.5% sodium nitrate/20-40% PEG in 10% sucrose solution), incubated at 35°C for 5 min followed by centrifugation at 200 g for 5 min, to obtain dense pellet. Protoplasts are transferred to a water bath at 30°C for about 30 min during which most of the protoplasts undergo fusion. The protoplasts are gently washed and transferred to microscope or plated. The fusion and fusion products are observed under inverted microscope - synthesis of cell wall, first cell division, formation of homokaryons/heterokaryons, frequency of multinucleate homokaryons from heterokaryons and mitotic index, cytological examination, etc are performed. Fig 4.1 outlines the various stages involved in the procedures for protoplast isolation.

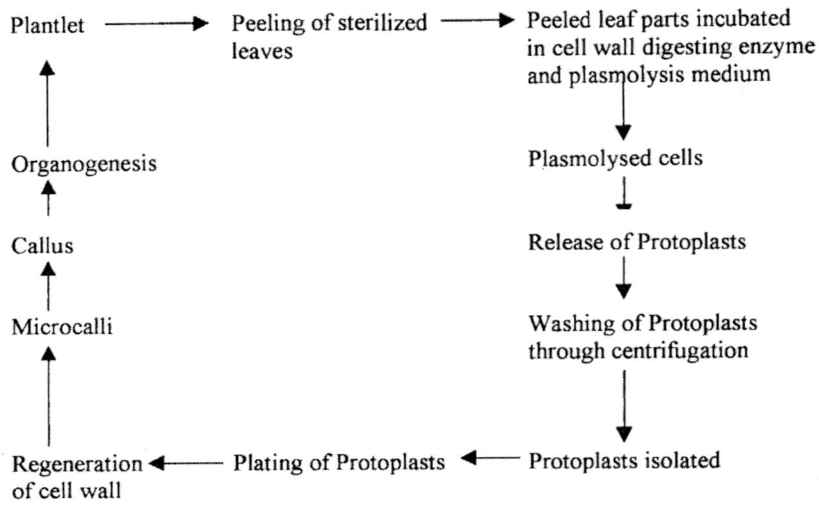

Fig. 4.1. Schematic representation of procedures in protoplast isolation

Methods Involved

(i) *Culture*: Protoplasts are grown and induced to undergo division in both liquid and agar medium. In solid media usually agarose is used as the coagulating agent.

(ii) *Plating*: Plating of the protoplasts allows to study the whole sequence of division, growth and development of a plant from single protoplast, and the plating efficiency (percentage of protoplasts forming colonies) is estimated.

(iii) *Fusion*: Two ml suspension of protoplasts (density 4×10^5) suspended in the liquid culture medium are poured into petri plates and mixed gently with an equal volume of the same medium containing 1 - 2% agar. Plates are then sealed with parafilm. The plates are incubated in an inverted portion at 28°C under continuous illumination from fluorescent tubes at 2200 lux.

(iv) *Cell wall*: The isolated protoplasts regenerate new walls around themselves. This can be demonstrated by staining the reconstituted protoplasts for 5 minutes with a fluorescent dye calcofluor white (0.1%).

(v) *Media*: The media for protoplast culture remain the same as used for culturing cells of that species, with minor changes.

(vi) *Division*: Protoplasts divide before regenerating cell wall. Again cells divide continuously to form an aggregate of cells, or a mass of callus tissue.

Regeneration from Protoplasts

Regeneration from protoplasts is through callus formation. The organogenesis or embryogenesis is achieved from callus cultures of the given species. In some cases, protoplasts can directly form somatic embryos e.g. in *Medicago, Brassica juncea* and *Citrus*. Following factors determine the chances of success in regenerating a plant from protoplasts (Binding *et al.*, 1982):

(i) The family, the species and the genotype.

(ii) Juvenile cells and cells from shoot tips have greater regenerative potential.

(iii) Use of rich media. Formation of callus from isolated protoplasts is often induced on Kao and Michayluk medium (1975), whereas the formation of embryos and organs is favoured by B5 medium with addition of 5-10% coconut milk.

The successful regeneration from protoplasts has been reported in several species (Table 4.3).

Fusogen Agents Employed in protoplast fusion

The earliest success was achieved using sodium nitrate (Carlson *et al.*, 1972). Different methods for protoplast fusion are:

(i) Spontaneous fusion of protoplasts: This can occur in mechanically or enzymatically isolated protoplasts.

(ii) Mechanically induced fusion: Through use of micromanipulators and micropippettes. e.g. as applied in *Glycine max, Vinca rosea*.

(iii) Sodium nitrate induced fusion.

(iv) High pH and Ca^{2+} induced fusion - pH of 10.5 and 50 mmol l^{-1} $CaCl_2.2H_2O$.

(v) Polyethyleneglycol (PEG) induced fusion (Kao and Michayluck, 1974). PEG is used for this purpose owing to high frequency of heterokaryon formation and low cytotoxicity to protoplasts. Dimethyl sulphoxide (DMSO) in combination with PEG stimulates protoplast fusion.

(vi) Dextran

(vii) Electric Fusion/Electroporation: (Zimmerman, 1982): Kameya (1984) emphasized the fusion of protoplasts using dextran and electrical stimulus. Here, protoplasts are brought in close contact by a non-uniform alternating current field between the two electrodes. Fusion is initiated by field pulse of high voltage (500 V - 3 kV, cm^{-1}) for 10-1000 microseconds. Protoplasts, when initiated into a high frequency alternating electric field (0.1-1 MHz) in a low conducting medium (e.g. mannitol) between parallel electrodes (0.1 - 1 mm apart), act as dipoles and are attracted to each other along strongest field lines. Once protoplasts come in contact, a high frequency of fusion is achieved by disrupting the plant membrane of protoplasts with a short D.C. pulse. This approach leads to production of viable heterokaryon. Once heterokaryon is formed, following events occur in order: wall regeneration by heterokaryotic cells, fusion of nuclei of heterokaryon to produce hybrid cell, plating and production of colonies of hybrid cells, selection of hybrid, sub-culture and induction of organogenesis in hybrid colonies, transfer of mature plants from regenerated callus.

Table 4.3. Examples of plants regenerated from protoplasts

Plants	Plant source	Regeneration
O. sativa var. *japonica*	Embryogenic cell susp. (ECS)	Plant
T. aestivum	ECS	Plantlet
H. vulgara	ECS	Albino plant
B. Oleracea var. *italica*	Hypocotyl	Plant
Cucumis sativus	Cotyledon & leaf	Plant
Asparagus officinalis	Embryogenic callus	Plant
Ipomoea batata	Petiole & stem	Plant
Beta vulgaris	Leaf	Plant
B. napus var. *oleifera*	Leaf	Plant
Helianthus annuus	Hypocotyl	Fertile plant
Picea glauca	ECS	Plantlet
Populus sp.	Leaf	Plant
Prunus avium	Cell susp. (root)	Plant
Rosa sp.	ECS	Plantlet
Pelargonium sp.	Cell susp. (Leaf, Petiole)	Plant
Chrysanthemum sp.	Cell susp. (Leaf)	Plant

Fusion Products

The nuclei of two protoplasts may or may not fuse together even after fusion of cytoplasm. The binucleate cells are known as heterokaryon. The nuclei when get fused, the cells are called hybrid/synkaryocyte; and when only cytoplasm fuse and genetic information from one of the two nuclei is lost, it is called cybrid/cytoplasmic hybrids/ heteroplast (Fig. 4.2) (Dodds and Roberts, 1985). There are certain genetic factors coded in cytoplasmic inheritance, instead of nuclear genes, e.g. male sterility in some plants. Susceptibility and resistance to some of pathotoxins and drug are under the control of cytoplasmic genes. Thus, cybrids have applications in plant breeding experiments (Vasil, 1982).

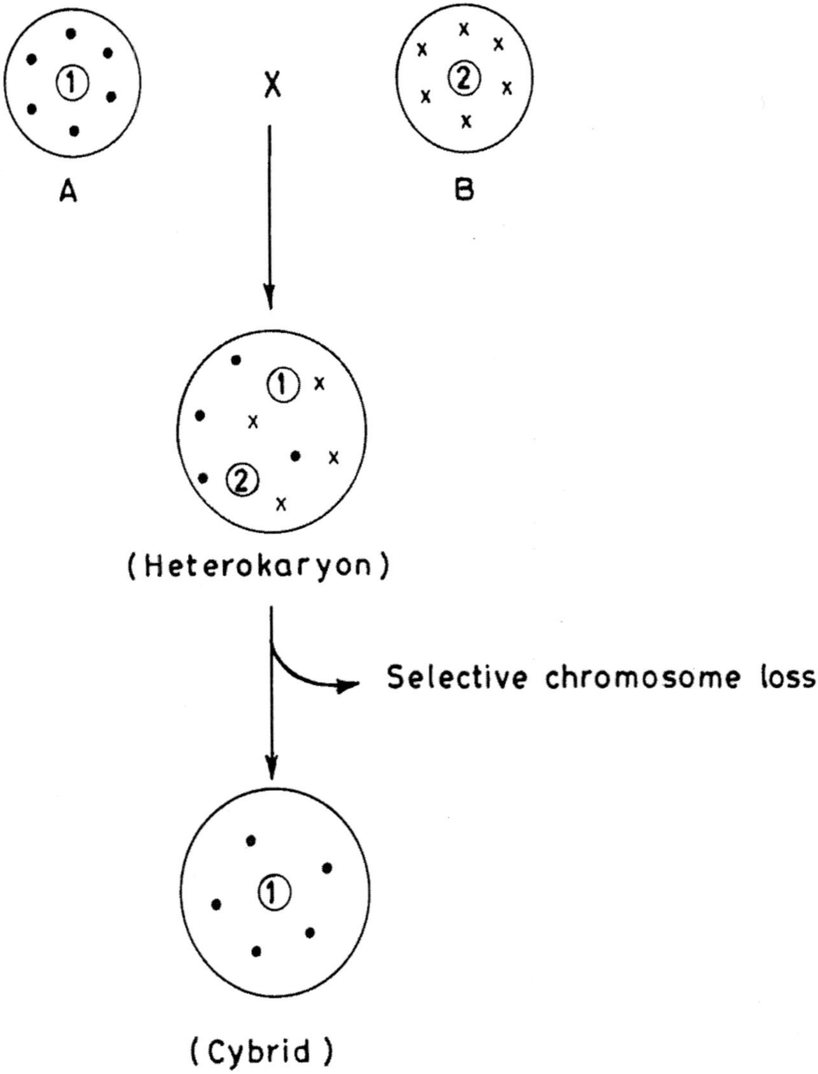

Fig. 4.2. Cybrid formation

Selection procedures after protoplast products/ somatic hybridization

To isolate the fusion products and for selection of plant cells for desirable traits, various strategies have been developed. e.g. preferential growth of somatic hybrids, mechanical selection, albino complementation, resistance to anti-metabolites, nutritional complementation, metabolic inhibitors, fluorescence activated cell sorting. Some of these selection methods are described as under:

(1) Direct isolation (Mechanical selection): It involves selection of morphologically different types of protoplast and isolation of heteroplasmic fused product under microscope by micro-manipulator or micropipette, and its culturing in special enriched media. This method was successfully employed to isolate hybrid resulting from fusion of mesophyll protoplast of *Nicotiana glauca* and cell culture protoplast of soybean.

(2) Resistance to anti-metabolites: Mutants resistant to amino acids, antibiotics or herbicides have been employed as dominant genetic markers for the selection of somatic hybrid (Harms, 1985). Cell selectable markers like drugs (Kanamycin or hygromycin) can be usefully utilized to improve efficient selection of somatic hybrid (Horsman *et al.*, 1997).

(3) Metabolic inhibitors: Somatic hybrids may also be selected using metabolic inhibitor alone or in combination with genetic markers. Iodoacetamide inhibits *Brassica oleracea* protoplast and poor regeneration ability of *B. campestris* has been used to select hybrid product with about 90% success.

(4) Fluorescence activated cell sorting: In this method two types of protoplasts are coloured with different fluorescent dyes (Galbraith, 1989). With the help of cell sorters the heterokaryons which contain both the dyes can be selected. The parental protoplast is marked by

staining with different vital fluorsecence dyes and the hybrid cell could be identified as it will have fluorescence characteristics of both the dyes. The double fluorescence characteristic of the somatic hybrid has been used to separate the fusion products from the parental protoplast using fluorescence activated cell sorter.

Application of Protoplast fusion and somatic hybridization

(1) Hybrid formation: Using somatic hybridization, it is possible to obtain **hybrids** which would not be possible by normal crossings, due to taxonomic or other barriers. e.g. somatic hybridization between *Solanum tuberosum* (tuber forming, little or no disease resistance) and *S. brevidens* (non-tuber forming, disease resistant).

(2) Formation of Cybrids: The genetic consequences of fusion hybrids show formation of three types of somatic hybrids:

(i) *Symmetric hybrids*: When normal protoplast of two species are fused, the resulting somatic hybrid plants have somatic chromosome complement of both the fusion parents. Such somatic hybrids are called **symmetric hybrids**. This phenomenon results in a new species. e.g. *Brassica naponigra* was produced by fusion of protoplast of *B. napus* and *B. nigra*.

(ii) *Asymmetric hybrids*: Hybrid cells can be obtained between distantly related taxa. These result due to preferential chromosome elimination of one of the fusion parents from normal somatic hybrid, either spontaneously or in response to an irradiation treatment. Unlike in symmetrical fusions, in asymmetric ones only a few chromosomes or tiny fragments of the chromosome are transferred from the donor species to the recipient partner. This strategy circumvents somatic incompatibility and sterility problems associated with plants derived from symmetrical fusions. Useful gene transfer during *in vitro* culture through **aymmetric hybrids** is possible.

(iii) *Cybrids*: These are obtained when nucleus is derived from one parent and cytoplasm is derived from both the parents (Fig 4.2). **Cybrids** can be obtained by using any one of the following methods:

(a) Fusion of normal protoplast from one parent with enucleated protoplast from the other parent.

(b) Fusion of normal protoplast from one parent and protoplast containing non-viable nuclei from the other parent.

(c) Selective elimination of one of the nuclei from the heterokaryon.

(d) Selective elimination of chromosome of one parent at a later stage after fusion of the nuclei.

Cybrids have been produced in many cases mainly to transfer cytoplasmic organelles. Cybridization has been successfully employed to make intergeneric and interspecific transfer of cytoplasm in tobacco, petunia, rice and *Brassica sp*.

(3) Direct DNA and Macromolecules Uptake: This is possible using PEG or electroporation. Polystrene, protein, DNA, viruses, bacteria or chloroplasts and nuclei can be transplanted into protoplasts. This system has been used to demonstrate mechanism of infection and nuclei protein replication. Transgenic plants are produced through DNA uptake into protoplasts. Some of the examples where plants have been regenerated using these methods are given in Table 4.4.

Table 4.4. Some examples of plant regeneration through direct DNA and macromolecule uptake in the protoplasts.

Species	Promoter/gene	Method employed	Regeneration
Oryza sativa	Promoter and the first intron of the alcohol dehydrogenase gene of maize/p-glucuronidase gene (GUS)	PEG	Plantlet
Solanum tuberosum	Nopaline synthase gene of *A. fumefaciens*/ neomycin	PEG	Plantlet
Arabidopsis thaliana	35s/hph	PEG	Fertile Plant
Lactuca sativa	35s/neo	Electroporation	Fertile Plant

(4) Some other examples of successful demonstration of protoplast fusion and somatic hybridization are given below (Also refer to table 4.5):

(i) protoplast fusion of *Nicotiana glauca* x *N. langsdorfii* (Carlson *et al.*, 1972).

(ii) somatic hybrids through interspecific cross of wild and cultivated variety of potatoes i.e., *Solanum tuberosum* x *S. chacoense* (Kuchko, 1985)

(iii) intergeneric hybrids which could be difficult to obtain by sexual crosses have been obtained using protoplast fusion. e.g. tomato + potato, *Datura* + *Atropa*, barley + rice, wheat + barley, wheat + oat.

(iv) potato + tomato (pomato) (Melchers *et al.*, 1978).

(v) inter-familial cell hybrids. e.g. *Glycine* sp. + *Nicotiana* sp. (Kao, 1977; Chien *et al.*, 1982).

(vi) production of fertile interspecific, diploid rice hybrid plants (Waara *et al.*, 1995).

(vii) a member of *Brassica*, *Moricandia arvensis* which is resistant to *Albugo candida*, *Alternaria brassicae* was crossed with *B. napus* using somatic hybridization to make *B. napus* disease resistant (Neil *et al.*, 1996).

(viii) Somatic hybridization is a promising method for citrus breeding programmes, for it provides an opportunity to overcome sexual incompatibility (Miranda *et al.*, 1997). The resulting somatic hybrid plant acts as a model for production of intergeneric somatic hybrid plants between citrus and related genera. Somatic hybrid plants are designated and registered as citrus parental lines by the Ministry of Agriculture, Forestry and Fishery, Japan for further breeding programmes (Nito *et al.*, 1998).

Assessment of somatic hybrid and cybrid nature of plants

To assess the hybrid nature of selected putative hybrid, some of the commonly employed methods are as follows:

(1) Morphological traits: Morphological selectable marker forms the basis. Intermediate expression of numerous vegetative and floral characters have been employed to identify hybrids. Identification of somatic hybrid between diploid clone of *Solanum tuberosum* and rol c transgenic diploid *S. papita* was possible using rol c gene as a morphological marker. The true somatic hybrid exhibited rol c phenotype i.e., reduced apical dominance with large number of adventitious shoots and a pale green colour of leaves (Kaendler *et al.*, 1996). This approach is not reliable as variations induced in hybrids by tissue culture may show new traits not displayed by either of the parents.

(2) Cytological studies: Somatic hybrids can be assessed on the basis of chromosome number, however, it may not

be applicable to all species, specially where fusion involves closely related species.

(3) Isozyme analysis: Isozymes are multiple molecular forms of an enzyme showing identical catalytic characteristics. If the two parents display different band patterns for a particular isozyme, the putative hybrid can be easily identified. On the basis of isozyme analysis, somatic hybrid between *S. napus* and *Moricandia arvensis* was identified (Neil *et al.*, 1996).

(4) DNA analysis: This approach serves as tool for analysis of somatic hybrids and cybrids. The proof of hybridity comes from the evidence of DNA from both parents.

(a) RAPD: Baird *et al.* (1992) used RAPD markers to identify both inter and intra-specific somatic hybrids of potato. RAPD method could be usefully applied at the early microcalli stage.

(b) RFLP: RFLP analysis of DNA is widely used to verify hybrid and cybrid nature of the selected fusion products. The presence of high density RFLP maps in tomato and potato has made it possible to determine the extent and direction of elimination in several asymmetric hybrids.

(c) SSR: Simple Sequence Repeat polymorphism is employed to analyse somatic hybrids, particularly at the callus level. It helps in characterizing somatic hybrid at both the inter- and intra-specific level.

Some examples signifying contribution of somatic hybridization in genetic improvement

(A) Cereals

Rice: Plants have been regenerated from isolated protoplasts of Japonica and Indica types of rice, thereby eliminating one of the major barriers to *in vitro* genetic manipulation of this cereal crop (Sink *et al.*, 1992).

Attempts have been made to multiply the conventional breeding barrier by interspecific fusion of rice with four different species including *Oryza brachyantha, O. eichingeri, O. officinalis* and *O. perrieri*. The production of fertile interspecific diploid rice hybrid plants have been possible through somatic hybridization (Waara *et al.*, 1995).

(B) Oil Seeds

(i) *Brassica* sp: *Moricandia arvensis* resistant to *Albugo candida, Alternaria brassicae* and *Plasmodiophora brassicae* was crossed with *Brassica napus* using somatic hybridization technology to make *B. napus* disease resistant (Neil *et al.*, 1996). The methodology used to enhance the level of industrially useful fatty acid in *Brassica napus* was based on exploitation of somatic hybridization using wild crucifer germplasm as fusion partner (Millam *et al.*, 1995).

(ii) *Helianthus annuus*: The technique of somatic hybridization has been tried to overcome the sexual incompatibility barriers.

(C) Vegetables

(i) *Lycopersicon esculentum*: The first somatic hybridization produced with tomato was the tomato + potato hybrid (Melchers *et al.*, 1978). The wild species have served as important source of agronomic traits via sexual process (Waara *et al.,* 1995). Many sexual crosses between tomato and wild species that were troublesome to perform due to bilateral and unilateral incompatibilty were overcome by somatic hybridization.

(ii) *Solanum melangena*: Sexual crossing barriers in egg plants have limit the transfer of agronomically important traits such as insect resistance and disease resistance, from related wild species. Using somatic hybridization desired properties have been incorporated (Waara *et al.,* 1995). Symmetric somatic hybridization between *S. melangena* and *S. khasianum, S. aethiopicum* etc have been produced.

(iii) *Solanum tuberosum*: Potato protoclones resistant to several pathogens like *Alternaria, Phytophthora, Fusarium* have been reported. *S. brevidens* is used as a fusion partner and hybrids expressing late blight and potato leaf roll virus resistance gene are reported. Frost tolerence and cold hardening traits have been transferred from *S. commersonii* to *S. tuberosum* by somatic hybridization (Millam *et al.*, 1995). Both interspecific (*S. tuberosum* + *S. brevidens*) and intergeneric (*S. tuberosum* + *Nicotiana tobaccum*) somatic hybrids have been obtained.

Fruits

(i) *Citrus* sp: Somatic hybridization aids in citrus breeding programmes providing an opportunity to overcome sexual incompatibility and to increase the variability of cytoplasmic genome which is maternally transmitted in sexual cross (Miranda *et al.*, 1997). It has resulted in production of intergeneric somatic hybrid plants between citrus and related genera. Functional mature fruit bearing citrus trees have been regenerated from protoplast fusion.

(ii) *Cucumis melo:* Cucurbits suffer economic loss due to different diseases. Resistance could not be transferred to melon by sexual crossing due to crossing barriers. Hence, asymmetric protoplast fusion using a melon and cucumber presents a way to transfer genes (Jarl *et al.*, 1995).

(iii) *Prunus* sp.: Somatic hybrids of *Prunus spinosa* and *Prunus cerasifera*, and hexaploid somatic hybrid have been produced (Ochatt *et al.*, 1995).

Ornamentals

(i) *Rosa hybrida*: The sexual barriers hampering the production of new cultivars have been overcome by somatic cell technologies through genetic recombination involving sexually incompatible genotype (Marchant *et al.*, 1997).

(ii) *Gentiana* sp.: *Gentiana* is an important ornamental plant in Japan. Its breeding is possible by intra- and interspecific crosses using only two species – *G. triflora* and *G. scabra*. The genetic base of common Gentiana cultivars, therefore, is rather restricted. Somatic hybridization can help improve the genetic base of ornamental plants.

Others

(i) *Nicotiana* sp.: In order to overcome sterility of sexual and somatic hybrid, asymmetric hybridization helps produce two asymmetric somatic hybrids with TMV resistance and female fertility properties. This enabled them to be back crossed to tobacco (Waara *et al.*, 1995).

(ii) *Medicago sativa*: Flowering somatic hybrids are produced between sexually compatible species, *M. sativa* and *M. falcata*. There are at least 20 reports on plant regeneration from protoplasts of 9 *Medicago* species, involving 14 reports on

M. sativa cultivars (lucerne, alfalfa). A large number of lucerne cultivars, breeding lines and clones have been identified capable of regeneration from callus tissue (Brown & Atanassov, 1985; Bingham *et al.*, 1988). Regeneration of plants was possible through protoplasts isolated from leaves, cotyledons, roots and suspension cells.

Advantages of Protoplast fusion and Somatic hybridization

(1) These techniques can be used as an alternative to obtain tetraploids and amphidiploids, when mitotic doubling is impossible.

(2) It makes possible the hybridization between plants that are still in juvenile phase and those already in flower.

(3) It allows genetic recombination of sexually incompatible genotype, which is impossible using conventional tehniques of hybridization.

Table 4.5. Examples of selected hybrids showing potential characters useful in crop cultivation

Plants	Traits transferred
(A) **Inter-specific hybrids**	
1. *Solanum pinnatisectum* + *S. tuberosum*	Late blight resistance
2. *Brassica napus* + *B. nigra*	Black leg resistance
3. *Cucumis melo* + *C. sativus*	Disease resistance
4. *Nicotiana tabacum* + *N. rustica*	Elevated nicotine content and resistance to black root rot disease
(B) **Inter-generic hybrids**	
1. *Brassica napus* + *Raphanus sativus*	Beet cyst nematode resistance
2. *B. napus* + *Moricandia arvensis*	Resistance to fungal disease; kranz leaf anatomy

3. *Citrus reticulata* + *Citropsis gilletiana*	Overcomes barriers to sexual hybridization
4. *Medicago sativa* + *Onobrychis vicifolia*	Bloat safe hybrid
(C) **Inter-familial hybrids**	
1. *Nicotiana tabacum* + *Daucus carota*	Increased genetic variability and overcoming sexual incompatibility
2. *Oryza sativa* + *Daucus carota*	
(D) **Cybrids**	
1. *Nicotiana tabacum* + *N. sylvestris*	Streptomycin resistance
2. *Brassica oleracea* + *B. napus*	Atrazine resistance

(4) Somatic hybridization helps to produce hybrids between species that are too distant to be hybridized sexually.

(5) The wild relatives of cultivated variety having dominant and resistance genes can be exploited to improve crop yield. For instance, wild relative of cultivated sweet potato contains dominant and resistant genes against root knot nematode.

(6) Both parental species equally contribute to hybrid in case of somatic hybridization.

(7) Protoplast can be pre-treated prior to fusion, which increases gene introgression.

(8) Asymmetric somatic hybridization helps eliminate chromosomes from hybrid cells and as a result fertile hybrids are regenerated. Thus, somatic incompatibility is by-passed in case of asymmetric somatic hybridization.

(9) Somatic hybridization makes possible transfer of polygenetic traits without knowing the molecular basis of genes coding the traits. It is of importance because many agronomic traits are polygenically controlled.e.g. transfer of high tuber yield genes in potato (Mattheij & Puite, 1992).

(10) It can be employed in the improvement of plants wherein sexual reproduction is either difficult or absent such as banana, cassava, potato, sugarcane and yam (Bhojwani & Razdan, 1996). Here, the fusion-derived plants need not be fertile, circumventing the problem of infertility.

Limitations (Disadvantages) of Somatic hybridization

(1) Non - availability of a definite marker which could identify somatic hybrids. RAPD markers were reported to identify both inter- and intra-specific somatic hybrid of potato (Baird *et al.*, 1992). The major drawbacks of using this for such analysis include troubles with reliability and reproducibility, co-migration of heterologous bands on agarose gel leading to mileading results, and the low level of polymorphism associated with these dominant markers (Provan *et al.*, 1996).

(2) The development of chimeric calluses, in place of hybrids; this results due to nuclei not fusing after cell division, and dividing separately.

(3) The end-products after somatic hybridization are sometimes unbalanced (low fertility/sterile/unstable) and so are not viable, particularly if the fusion partners are taxonomically far apart. e.g. most somatic hybrid of *Solanum tuberosum* + *S. commersonii* did not produce normal pollen grains i.e., had reduced pollen fertility (Thieme *et al.*, 1997).

(4) The genetic stability during protoplast, cell and callus culture is poor.

(5) Regeneration products produced following somatic hybridization sometimes are very variable, which could be on account of (Harms, 1983):

(i) translocation (ii) somaclonal variation (iii) chromosome elimination, resulting in aneuploids, for instance, and (iv) organelle segregation.

SUGGESTED READINGS

Baird, E., Cooper-Bland, S., Waugh, R., De Maine, M. & Powell, W. (1992). *Mol. Gen. Genet.*, **233**, 469-475.

Bhojwani, S.S. & Rajdan, M.K. (1996). Plant tissue culture: theory and practice. Elsevier Publ., Amsterdam, p. 503.

Binding *et al.* (1982). *In: Plant Tissue Culture* (ed. Fujiwara). Proc. 5th Intl. Congr. Plant Tissue Cell Culture, Tokyo, Japan, pp. 575-578.

Bingham, E. T., McCoy, T. J. & Walker, K. A. (1988). *In: Alfalfa and alfalfa improvement* (ed. Hanson, A. A.). Agronomy monograph vol 29, ASA-CSSA-SSSA, Madison, USA, pp. 809-826.

Brown, D.C.W. & Atanassov, A. (1985). *Plant Cell Tissue Org Cult.*, **4**, 111-122.

Brown, D.C.W. & Thorpe, T.A. (1995). *World J Microbiol. Biotech.*, **11**, 409-415.

Carlson, P.S., Smith, H.H. & Dearing, R.D. (1972). *Proc. Nat. Acad. Sci.,* USA, **69**, 2292-2294.

Chien, Y.C., Kao, K.N. & Wetter, L.E. (1982). *Theor. Appl. Genet.*, **62**, 301-304.

Dodds, J. H. & Roberts, L. W. (1985). *Experiments in Plant tissue culture,* 2nd Ed. Cambridge University Press, Cambridge.

Fromm, M.E., Taylor, L.P. & Walbot, Y. (1985). *Proc. Nat. Acad. Sci.,* USA, **82**, 5804-5828.

Galbraith, D.W. (1989). *In: Biotechnology in agriculture and forestry. Vol 9. Plant protoplasts and genetic engineering. II* (ed. Bajaj, Y.P.S.). Springer Veralg, Berlin, pp 304-327.

Harkins, K.R. & Galbraith, D.W. (1984). *Physiol. Plant.*, **60**, 43-52.

Harms, C.T. (1985). *Quart. Rev. Biol.*, **58**, 325-353.

Harms, C.T. (1985). *In: Plant protoplast* (ed. Fowke, L.C. & Constable, F.). CRC Boca Raton, pp 169-203.

Horsman, K., Bergervoet, J.E.M. & Jacobsen, E. (1997). *Euphytica*, **96**, 345-352.

Jarl, C.I., Bakelmann, G.S. & Deltaas, J.M. (1995). *Plant Cell Tissue Org Cult.*, **43**(3), 259-265.

Kameya, T. (1984). *In: Cell culture and somatic cell genetics. Vol 1. Laboratory procedures and their applications* (ed. Vasil, I. K.). Orlando, Florida, pp 423-427.

Kaendler, C., Fladung, M. & Uhrig, M. (1996). *Applied Genet.*, **92** (3-4), 455-462.

Kao, K.N. (1977). *Mol. Gen. Genet.*, **150**, 225-230.

Kao, K.N. & Michayluk, M. R. (1975). *Planta*, **126**, 105-110.

Kartha, K.K., Gamborg, O.L., Constable, F. & Shyluk, J.P. (1974). *Plant Sci. Lett.*, **2**, 107-113.

Klercker, J.A. (1892). *Oefvers Vetenskaps Adad.*, Stockholm, **9**, 463-471.

Krens, F.A., Molandijk, L., Wullems, G.J. & Schilperoort, R.A. (1982). *Nature*, **296**, 72-74.

Kuchko, A. A. (1985). *In: Plant cell culture* (ed. butenko, R. G.). MIR, Moscow, pp 154-174.

Kuster, E. (1909). *Ber. Dtsch. Bot. Ges.*, **27**, 589-598.

Marchant, R., Daney, M.R. & Power, J.B. (1997). *Plant Cell Tissue Org Cult.*, **50**, 131-134.

Mattheij, W.M. & Puite, K.J. (1992). *Theor. Appl. Genet.*, **83**, 807-812.

Melchers, G., Sacristan, M.D. & Holder, A.A. (1978). *Carlsberg Res. Commun.*, **43**, 203-218.

Millam, S., Payne, L.A. & Mackay, G. R. (1995). *Euphytica*, **85**(1-3), 451-455.

Miranda, M., Motomura, T., Ikeda, F., Ohgawara, T., Saito, W., Endo, T., Omura, M. & Moriguchi, T. (1997). *Plant Cell Rep.*, **16**, 401-405.

Neil, N., Murata, T., Morgan, C.L. & Mathias, R.J. (1996). *Theor. Appl. Genet.*, **93**, 1234-1241.

Nito, N., Katayama, Y. & Yamaguchi, S. (1998). *In: Crop productivity and sustainability - Shaping the Future.* Proc. 2nd Intl. Crop Sci. Congr. (eds. Chopra, V. L., Singh, R. B. & Varma, A.). Oxford & IBH Publ., pp 383-393.

Ochatt, J. Serigo & Estela, P.O. (1995). *Euphytica*, **85**, 287-294.

Provan, J., Kumar, A., Shepherd, L., Powell, W. & waugh, R. (1996). *Plant Cell Rep.,* **16**, 196-199.

Schweiger, H.G., Dirk, J., Koop, H.U., Kranz, E., Neuhaus, G., Spangenberg, G. & Wolff, D. (1987). *Theor. Appl. Genet.*, **73**, 769-783.

Sink, K.C., Jain, R.K. & Chowdhury, J.B. (1992). *In: Distant hybridzation of crop plants* (eds. Kallo, G. & Chowdhury, J. B.). Springer-Verlag, NY, pp. 168-198.

Takabe, I., Otsuki, Y. & Aoki, S. (1968). *Plant Cell Physiol.*, **9**, 115-124.

Thieme, R.D., Ulrich, G., Tatyana, D.D. & Tiemamn, H. (1997). *Euphytica*, **97**, 189-200.

Waara, S. & Glimelius, K. (1995). *Euphytica*, **85**, 217-233.

Zelcer, A., Aviv, D. & Galun, E. (1978). *Z. Pflanzenphysiol.*, **90**, 397-407.

Zimmerman, R.H. (1982). *Biochim. Biophys. Acta*, **694**, 227-277.

QUESTIONS

1. Describe methods of isolating protoplasts from plant cells.
2. Comment on fusion methods and regeneration from protoplasts.
3. Discuss applications of protoplast culture and somatic hybridization giving examples.
4. Write short notes on the followings:
   (1) Protoplasts viability test
   (2) Factors governing protoplast culture
   (3) Fusogen agents
   (4) Hybrids and cybrids.

CHAPTER - 5

SYNTHETIC SEEDS OR SOMATIC SEEDS OR ENCAPSULATED SEEDS

Synthetic seeds are defined as artificially encapsulated somatic embryos, shoot buds, cell aggregates, or any other tissue that can be used for sowing as a seed and which possess the capability to convert into a plant under *in vitro* conditions. Somatic embryogenesis, organogenesis and axillary bud proliferation systems are excellent techniques for rapid and mass scale production of synthetic seeds *in vitro*. Encapsulation is expected to be the best method to provide protection and to convert the *in vitro* derived propagules into 'synthetic seeds' or 'synseeds' or 'artificial seeds' (Gray, 1986; Redenbaugh *et al.*, 1986, 1993; Ara *et al.*, 2000). The synthetic seed technology combines the advantages of clonal propagation with those of seed propagation and storage.

Types of synthetic seeds

(i) hydrated synthetic seeds: consist of somatic embryos individually encapsulated in hydrogel such as calcium alginate.

(ii) desiccated synthetic seeds are produced by coating somatic embryos in polyoxyethylene glycol.

Synthetic seed is a newer concept in seed biotechnological research and is mainly considered for propagation and delivery of tissue cultured plants in a more economical and easy way. Hydrated encapsulation of somatic embryos/propagules from recalcitrant crops has proved useful and provides a method for increasing the short storage life of the seeds (Redenbaugh *et al.*, 1991). For bio-reaction development and standardizing systems for synchronously developing propagules, followed by automation of the whole process of sorting, harvesting, encapsulation and germination of the coated propagules, can speed up production of synthetic seeds.

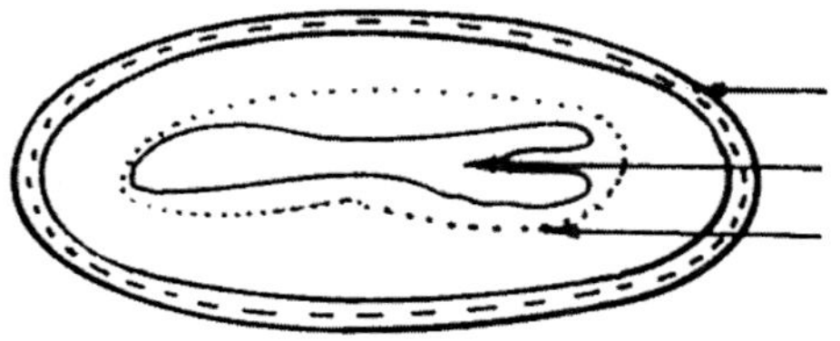

Fig. 5.1. An encapsulated seed of plant

T. Murashige (1977) gave the concept of artificial seeds for the first time. Artificial (encapsulated) seeds are the somatic embryos covered with a protecting gel. These seeds are compared to the true seeds. In these seeds, the gel acts as seed coat and artificial endosperm providing nutrient as in true seeds (Fig. 5.1). Water-soluble gels (hydrogels) must be used as the protective gel. Usually Na/Ca alginate or carrageenan (a product of brown algae) is selected for encapsulation purpose because it is less toxic to embryos and easy to handle. There are several gels used for coating embryoids (Table 5.1). Shoot tips excised from the shoot cultures of banana have been encapsulated in 3% sodium alginate solution (Ganapathy *et al.*, 1992). Alginate artificial seeds are spherical, transparent and non-inhibitory. The production of 'seeds' by coating somatic embryos and getting plants from these protected embryos is far from easy because: (i) the development of these artificial seeds which are stable for many months needs physical and chemical methods for making embryo quiescent. (ii) such encapsulated seeds are to be protected against desiccation when stored under dry conditions.

Two systems are developed for handling of synthetic seeds:

(1) Encapsulation: A gel containing nutrients is coated around somatic embryo. An encapsulation machine is developed which can produce and sort around 6,000 single embryos per hour.

(2) Sowing the somatic embryos by fluid drilling (Gray, 1986).

Four types of synthetic seeds are proposed based on embryos and their encapsulation:

(a) Uncoated, desiccated somatic embryos e.g. for orchard grass.

(b) Coated, desiccated somatic embryos e.g. carrot.

(c) Encapsulated (coated), hydrated somatic embryos, e.g. alfalfa.

(d) Uncoated hydrated embryos (in a fluid drilling gel), e.g. carrot.

Methods of Production

Following steps are followed:

(1) Induction of pseudoembryos (artificial embryos) from cell suspension culture.

(2) Mixing of embryos with 2% Na-alginate.

(3) Immersing the embryos in a bath of calcium salt, e.g. solution of $Ca(NO_3)_2$ for 30 min. This leads to very quick complex formation at surfaces due to exchange of ions i.e., Na^+ and Ca^+. As a result, individual embryo is enclosed into a clear and hardened beads of about 4 mm. Hardening of calcium alginate is modulated with concentrations of sodium alginate along with the duration of complexing. Usually 100 mM Ca^{2+} is used.

(4) Sieving the bead through a nylon mesh; $Ca(NO_3)_2$ solution is recycled, and

(5) Testing the growth vigour of beads by plating in sand or soil amended with pesticides.

Injection of Somatic Embryos

↓

Maturation of Somatic Embryos

↓

Encapsulation of Somatic Embryos

↓

Evaluation of Embryoid and Plant Conversion

↓

Plating in fields/Green House

Thus synthetic seeds are living seed-like structure derived from somatic embryoids after encapsulation by a hydrogel. Such preserved embryoids are called as synthetic seeds. These encapsulated embryos can resist unfavorable field conditions including microbial contamination, without desiccation. Such encapsulated seeds are used as a substitute of natural seeds and can also be grown directly in the greenhouse or in the fields.

Table 5.1. Different gels for encapsulation of somatic embryos

Gels	Conc. % (w/v)	Complexing agent	Concentration (mM)
Sodium alginate	0.5 - 5.0	Calcium salts	30 - 100
Sodium alginate with gelatin	2.0	Calcium chloride	30 - 100
Carrageenan with	0.2 - 0.8	Potassium or	
Locust Bean Gum	0.4 - 1.0	Ammonium chloride	500
Gelrite	0.25	Low Temperature	

Kitto and Janick (1985) produced *Citrus* embryos *in vitro* and tested eight compounds for their synthetic coating properties on embryos. Out of these, a polyethyleneoxide (Polyox WSR-N 75) revealed good encapsulating properties, as shown with *in vitro* produced carrot embryos (Kitto and Janick, 1985). Examples of plants produced from synthetic seeds sown *in vitro* and in soil include: *Apium graveolens, Brassica* sp., *Gossypium hirsutum, Medicago sativa, Oryza sativa, Zea mays, Picea abies (in vitro*), and *Apium graveolens, Daucus carota, Medicago sativa.* (in soil).

In addition to the use of somatic embryos as propagules for encapsulation, axillary buds have also been successfully used in plants in which somatic embryogenesis is difficult to achieve, as in mulberry and other woody species (Bapat *et al.* 1987; Piccioni and Standardi, 1995). Apart from these unipolar structures (apical shoot tips and axillary shoot buds), apolar protocorms or protocorm-like bodies as well as undifferentiated embryogenic calli are also used in synthetic seed production (Table 5.2). Synthetic seeds are produced by encapsulating the protocorm or prorocorm-like bodies in sodium alginate gel, in case of orchids like *Cymbidium giganteum, Dendrobium, Phaius* and *Spathoglottis.* Production of synthetic seeds from apical shoot tips and axillary shoot buds is facilitated by treating these organs first with auxins for root formation and then the microcuttings (4 mm long) are encapsulated in sodium gel. However, no such treatment for root induction was done to encapsulated shoot buds for regenerating plants of mulberry (Bapat & Rao, 1990) and banana (Ganapathi *et al.*, 1992). Bacterial contamination is controlled by an antibiotic mixture (0.25 mg/l) containing cefatoxime (250 mg), rifampicin (60 mg) and tetracycline-HCL (25 mg) dissolved in 5 ml dimethyl sulphoxide (Ganapathi *et al.*, 1992). Table 5.2 lists important plant species in which synthetic seed technology has been successfully employed.

RECENT DEVELOPMENTS IN SYNTHETIC SEED TECHNOLOGY

(1) Automate encapsulation method

It is the process which in recent times fastens the production of artificial seeds. Encapsulation machine has been used successfully to encapsulate somatic embryos, as employed for alfalfa. In this method, blank alginate capsules are planted in speeding trays with the help of a vacuum seeder. These capsules are then put in the field using a Stanhay planter. To avoid rapid drying of these alginate capsules, Elvax 4260 co-polymer which acts as hydrophobic agent, is coated to help conserve embryo.

(2) Mass balance

This concept has improved the technology of synthetic seeds, which is based on the amount of tissue at the start of the experiment and the number of high quality plants produced at the end of the operation.

Commercial Production of Synthetic Seeds

Schematic representation of steps required to produce synthetic seeds on commercial scale is illustrated as under:

Production of embryogenic tissue from transformed tissue

↓

Mass production of somatic embryos and their maturation

↓

Non-inhibitory coating (encapsulation)

↓

Storage potentiality of these artificial seeds

↓

High frequency, direct nursery field conservation

↓

Low genetic and epigenetic variation

↓

Expression of engineered trait

Potential Uses of Artificial Seeds

(1) Reduced costs of transplants.

(2) Production of large number of identical embryos.

(3) Germplasm conservation through cryopreservation.

(4) Study of seed coat formation is possible.

(5) Usefulness in tree genetic engineering.

(6) Production of synthetic seeds possible any time, in any season of a year.

(7) Problem of dormancy can be avoided through synthetic seed technology, thereby shortening the life cycle of a plant.

(8) It is possible to undertake study related to anatomical features of endosperm and seed coat formation.

(9) On commercial front, the encapsulated embryos can be packed on large scale with suitable pesticides, nitrogen fixing microorganisms and fertilizers.

Table 5.2. Synthetic seeds in some important plant species

Plants	Propagules used for encapsulation
Actinidia deliciosa	shoot buds
Arachis hypogaea	somatic embryos
Asparagus cooperi	somatic embrtyos
Betula pendula	shoot buds
Brassica campestris	shoot buds
Camellia japonica L.	somatic embryos
Cymbidium sp.	protocorm-like bodies
Daucus carota	somatic embryos
Eucalyptus citriodora	somatic embryos
Mangifera indica	somatic embryos
Musa sp.	shoot buds
Pinus taeda	somatic embryos

Pistacia vera	embryogenic masses
Picea mariana	somatic embryos
Psidium guajava	somatic embryos
Rubus sp.	shoot buds
Santalum album	somatic embryos
Solanum melongena	somatic embryos
Syringa vulgaris	axillary buds
Vitis vinifera	somatic embryos
Zingiber officinale	shoot buds

Limitations of Synthetic Seed Technology

(1) Recovery of plants from encapsulated somatic embryos is often very low, due to various reasons:

- (i) Incomplete embryo formation
- (ii) Difficulties to arrest growth
- (iii) Difficulties in creating an artificial endosperm within the capsule. Hardening treatments like pre-treatment of embryogenic cultures with 12% sucrose, chilling and ABA have been described by Kitto & Janik (1985) to enhance the survival of encapsulated carrot embryos.

(2) Lack of dormancy and stress tolerance in somatic embryos limit the storage of synthetic seeds.

(3) The embryos are susceptible to microorganisms, so use of antibiotics and fungicides is recommended to check microbial contamination.

(4) Anomalous and asynchronous development of somatic embryos.

Maturation of somatic embryos which controls germination and conversion rate, remains a critical factor in synthetic seed technology. Addition of high sucrose content in the medium having ABA prevents maturation, whereas

PEG addition along with ABA improves frequency and synchrony of somatic embryo maturation (Fowke & Attree, 1996). Further, use of maltose and sucrose helps improve the viability and conversion rate of somatic embryos into plantlets. Another problem associated with synthetic seed technology is the diversity in the morphology of somatic embryos which eventually limits the use of somatic embryos. ABA treatment of somatic embryos leads to production of anomalous cotyledons (Lee and Soh, 1994); similarly, cytokinin treatment increases the number of somatic embryos with multiple cotyledons.

Desiccation damages the somatic embryos and inhibits their germination and conversion into plants. Besides, coating material also influences the success of the synthetic seed technology. Calcium alginate caoting is very wet and stick together, besides this, it loses water and dry down to a hard mass in a short time upon exposure to ambient atmosphere. The other criterion affecting the synthetic seed technolgy is the concentration of coating material. The coat should contain nutrients and growth regulator that are necessary for germination and conversion. In this regard, sodium alginate from brown algae is the best option.

SUGGESTED READINGS

Ara, H., Jaiswal, U. & Jaiswal, V.S. (2000). *Curr. Sci.*, **78**, 1438-1444.

Bapat, V.A. & Rao, P.S. (1990). *Plant Cell Tissue Org Cult.*, **20**, 69-70.

Bapat, V.A., Mhatre, M. & Rao, P.S. (1987). *Plant Cell Rep.*, **6**, 39-395.

Fowke, L. & Attree, S. (1996). *Plant Tissue Cult. Biotechnol.*, **2**, 124-130.

Ganapathi, T.R., Suprasanna, P., Bapat, V.A. & Rao, P.S. (1992). *Plant Cell Rep.*, **11**, 571-575.

Gray, D.J. (1986). *Hort. Sci.*, **21**, 820.

Kitto, S.C. & Janick, J. (1985). *Hort. Sci.*, **20**, 98-100.

Lee, K.S. & Soh, W.Y. (1994). *Korean J Plant Tissue Cult.*, **21**, 287-291.

Murashige, T. (1977). *Acta Hort.*, **78**, 17.

Piccioni, E. & Standardi, A. (1995). *Plant Cell Tissue Org. Cult.*, **42**, 221-226.

Redenbaugh, K. (1993). *Synseeds: Applications of synthetic seeds to crop improvement.* CRC Press, Boca Raton, USA.

Redenbaugh, K., Fujii, J., Slade, D., Viss, P. & Kossler, M. (1991). *In: Biotechnology in agriculture and forestry.* Vol 17 (ed. Bajaj, Y. P. S.). Springer Verlag, Berlin, pp 395-416.

Redenbaugh, K., Paash, B.D., Nichol, J.W., Kossler, M.E., Viss, P.R. & Walker, K.A. (1986). *Bio/Tech.*, **4**, 797-801.

QUESTIONS

1. What is synthetic or encapsulated seeds? Describe methods adapted to produce synthetic seeds.
2. Write short notes on the followings:

   (i) Uses and limitations of synthetic seed technology

   (ii) Automate encapsulation method.

CHAPTER - 6

PRODUCTION OF HAPLOIDS

Haploid plants have the gametophytic number of chromosomes, a single set (n) of chromosomes in the sporophyte. Haploid cells or individuals are those in which original chromosome number has been reduced by half. These are of immense importance for plant improvement, useful in mutations and for the production of homzygous plants. Conventional breeding programmes are tedious and are not very efficient for the production of homozygous haploids. These methods are more or less unsuccessful in cross-pollinated crops for obtaining homozygous plants.

Tulecke (1953) for the first time showed that mature pollen grains of *Ginkgo biloba* can be induced to proliferate in culture to produce haploid callus. This demonstrated that haploid callus can be produced from highly differentiated cells like pollen instead of normal process of pollen germination. Later on, Sipra Guha and Satish C. Maheshwari in 1964 reported the direct development of embryoids from microspores of *Datura innoxia* by the culture of excised anther. Bourgin and Nitsch (1967) obtained the first haploid plants from isolated anthers of *Nicotiana tabacum*.

The production and occurrence frequency of haploids *in vivo* can be improved by many techniques:

(1) Gynogenesis: Development of an unfertilized egg-cell, either as a result of delayed pollination, through use of

abortive pollen or through use of alien pollen. It is found, particularly, in inter-specific crosses.

(2) Androgenesis: The nucellus of egg cell is eliminated before fertilization, and the haploid individual is provided by the development of the egg-cell having male nucleus.

(3) Genome elimination by distant hybridization: This occurs due to intergeneric and interspecific crosses. After fertilization, one genome is eliminated giving rise to an embryo that has only one genome present, and is thus haploid. e.g. interspecific cross between *Hordeum vulgare* and *H. bulbosum* results in production of haploid *H. vulgare*.

(4) Chemical treatment: Maleic hydrazide, toluene blue, colchicine and nitrous oxide may produce haploids. Addition of chloramphenicol and p-fluorophenylalanine leads to chromosome elimination, and thus haploids are produced.

(5) Semigamy: The nucleus of egg-cell and the generative nucleus of the germinated pollen grain divide independently, resulting in a haploid chimera.

(6) Low or high temperature shock.

(7) Irradiation with X-rays or UV rays.

These *in vivo* methods show a low frequency of haploid production. Contrary to this, *in vitro* methods have produced haploids by anther culture or pollen culture, from about 250 species and hybrids. Members of the Solanaceae are particularly responsive and are capable of regenerating haploids from isolated anthers or individual pollen grains (Bhojwani *et al.*, 1986). Besides, in families of Cruciferae, Gramineae and Ranunculaceae different genera can also form haploids (Bajaj, 1983). Thus it is apparent that three pathways operate to induce haploid sporophyte of higher plants:

(a) Haploid sporophyte arising from meiotic spore through pollen (anther) and/or, ovary and ovule culture.

(b) Haploid sporophyte arising from zygote due to chromosome elimination.

(c) Haploid sporophyte arising from male and female gametes.

Procedure for haploid production *in vitro*

Closed flower buds, which have anthers containing uninucleate microspores, are most suitable for the induction of androgenesis. Excised flower buds are sterilized, cut open longitudinally and anthers are transferred on the medium. Filaments from anthers should be removed. Any injury to anthers should be checked during excision and transfer, which otherwise may very often produce callus from parts other than pollen. These anthers can be grown on solid or in liquid medium. The cultures are incubated at 25 - 27°C and under illumination of about 2000 lux for 12-16 h per day. Depending upon the plant species it takes certain time, about 1 - 8 weeks for plantlets to emerge from the anthers, after which these can be transferred directly into pots after necessary washing.

The viable gametes can only be formed in these haploid plants by getting homozygous diploids. Haploid plants can be diplodized to produce homozygous plants by two means: (1) by treating the young plantlets for 24-28 h with 0.5% aqueous colchicine solution while they are still attached to the anther. After washing they are transplanted to pots. Mature plants are treated with colchicine lanolin paste (0.4%) to obtain diploid callus shoots. This produces homozygous seeds. (2) Haploid stem segments can also be cultured. Such cultures tend to become diploid by endomitosis, thus diploid homozygous plants can be produced from such cultures.

Culture Media, Nutritional Requirements and other factors influencing anther cultures

Basal media of White (1934), Murashige and Skoog (1962) and Nitsch and Nitsch (1969) are used for cultures of excised anthers. Potato extract medium produces high yields in monocotyledons. A combination of plant growth regulators and suitable concentration of sugar (2 - 4%) are added to the media. High sucrose level (6 - 12%) is beneficial for androgenesis, as demonstrated in case of wheat, barley and tomato; sucrose acts as osmoticum rather than nutrient because anther produces small amount of tissues. The nutritional requirement of anther culture is simpler than that of isolated pollen grains. Isolated pollen grains are highly specialized cells and all the nutritional factors provided by anther wall are not present for isolated pollen grains. Therefore, pollen cultures may require additional nitrogen, particularly in the form of amino acids (organic reduced nitrogen source). L-glutamine is the most frequently used amino acid for haploid production from anther culture of potato, *Brassica napus* and *Triticum aestivum*. Proline, D-amino butyric acid, serine and Fe-EDTA induce embryony. In *Nicotiana tabacum*, addition of 2% charcoal in the medium increases the response of anthers from 41 to 91%. The pH is generally adjusted to 5.8. Culture of isolated microspore is shown in many crop plants like tobacco, rice, rapseed, barley, wheat, potato, etc.

Temperature treatment increases the induction frequency of microspore androgenesis. The usefulness of effects of cold/hot pretreatments to anthers/buds has been demonstrated in anther cultures of several species. The buds which are treated at 3°C or 5°C (cold treatment) for about 72 h induced approximately 58% anthers to yield pollen embryos in *Datura*, *Nicotiana*, as compared to 21% anthers from buds exposed to 22°C for same period (Nitsch, 1974). The beneficial effects of cold treatment are reported (Tsay *et al.*, 1988; Ogawa *et al.*, 1992). Such treatment depends on the cultivars. Reports are also available that suggest for pre-treatment of anther to elevated temperature (35°C) in order to stimulate androgenesis, in some *Brassica* and *Capsicum* species (Bhojwani & Bhatnagar, 1990). Physical factors such as day length, light quality, light/dark cycle etc) during the *in vitro* culture can also play an important role in androgenesis (Debergh, 1978; Bajaj, 1983; Dunwell, 1985).

The donor plants (starting materials) and their conditioning are vital for androgenesis. Healthy and strong plants should be used. The age of the donor plant, as well as age of the flower bud are important factors. The composition of the nutrient medium may need to be modified for different developmental stages during anther culture (Wenzel & Fouroughi-Wehr, 1985); for potato, certain components in the media need to be changed, and these include: macro-salts, sucrose, auxin, cytokinin, coconut milk and activated charcoal. In anther culture, anthers are grown directly over a solid medium. In pollen culture, the anthers are made to liberate the pollen in liquid medium and then plated on a solid medium applying standard techniques. Factors that affect plant growth like plant age, light intensity, photoperiod, temperature, season also affect the response of anther culture.

Isolated Anthers or Individual Pollen grains?

Most of the haploids which are obtained *in vitro* have arisen from isolated anthers and only sometimes from pollen. The growth of individual pollen grains and the regeneration from these haploids, have a number of advantages over anther culture (Sunderland, 1978; Bajaj, 1983):

(1) Chance of regeneration of diploids is limited by removal of diplod tissues like anther wall, tapetum.

(2) Anther does not act as a barrier to transport of nutrients from medium to the pollen grains.

(3) Pollen grains are more suitable than anthers as starting material for mutation research and genetic manipulation.

(4) Better embryo formation can be achieved with pollen grains.

(5) Callus formation from anther can be avoided, resulting in far fewer chimeras.

Despite these advantages of pollen culture, it is still very difficult to induce individual pollen grain to form haploid. A more complicated nutrient medium is needed than for anther culture, and still the number of haploids produced are far fewer. Pollen culture for haploid production is unsuccessful for most plant, but can be achieved in following species: *Petunia hybrida, Brassica oleracea, Nicotiana tabacum, Datura innoxia* and *Lycopersicon lycopersicum*. Major advances have been made in the development of isolated microspore culture technique (Keller, 1986).

Development of Androgenic haploids

There are many pathways of androgenesis which leads to the formation of haploid either directly by embryogenesis or indirectly through callus formation (Fig. 6.1). Direct embryogenesis is preferred over indirect embryogenesis involving callus formation which may produce variation. There are several hundred plant species in which haploid callus or complete plantlets have been produced. A few selected species are listed in the Table 6.1.

The pollen mother cell (P.M.C.) gives rise to pollen tetrads after meiotic division. These microspores are released in the atmosphere. The first mitotic division in the microspores gives rise to a large regenerative cell and small generative cell. The former remains quiescent while the later divides to produce two male nuclei. In culture although androgenesis can be induced in anthers in the tetrad stage at the binuclear pollen stage, but microspores just before or at the time of first mitosis, are most suitable for the induction of androgenesis. Examples of androgenesis are observed in datura, tobacco and barley. Androgenesis may be produced by pollen grains through mitosis producing equal or unequal nuclei. They produce somatic embryos directly or may produce callus. Direct androgenesis is evident in *Atropa belladona, Datura innoxia* and *Nicotiana tabacum*. Indirect androgenesis is the most common method (96%) of producing haploids. This type of development (indirect androgenesis) is supposed to be result of disturbed polarity and/or complex nutrient medium.

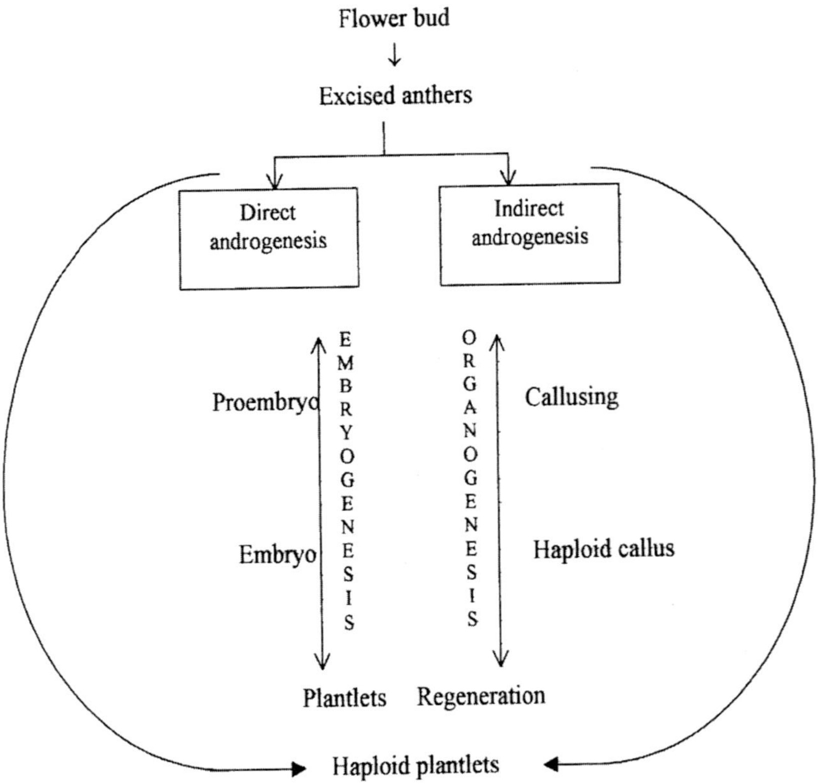

Fig. 6.1. Schematic presentation of haploid plant production

Table 6.1. Some examples of haploid plants obtained by excised anther/pollen

Plant Species	Inoculum Used	Response Observed
Asparagus sp.	Anther	Callus, plantlet
Brassica napus	Anther	Callus, embryos
Nicotiana tabacum	Anther	Callus, plants
Zea mays	Pollen	Embryos
Oryza sativa	Anther	Callus, embryos, plantlets
Coffea arabica	Anther	Callus

Microspore embryogenesis has improved in recent years in case of cereals, particularly wheat, rice and barley. Direct embryogenesis (no callus phase) remains the preferred choice/route in crop improvement programmes involving hybrids. Several factors affect the response of anther culture:

(i) The effect of genotype of donor plants i.e., genetic factor. e.g. major differences between different genotypes of wheat, rice and baeley have been observed with respect to yield of callus per anther cultured and regenerant frequency of green plantlets.

(ii) Microspore developmental stage: Early uninucleate, mid-uninucleate, late-uninucleate and pre-mitotic stages of microspore decide the response of anther culture. e.g. yield of callus declines when anthers bearing microspores at the stage other than mid – or late uninucleate stage are cultured.

(iii) Culture media: Anther culture of cereals requires modification in carbon and nitrogen components for improved anther culture. e.g. reduced concentration of ammonia nitrogen in the induction medium is effective. Sucrose, maltose and other monosaccharide as source of carbon have proved beneficial (Hunter, 1988; Chu *et al.*, 1990). Presence of 2,4-D influences the response of cultures: wheat anther culture showed positive response in presence of high amount (4 mg l^{-1}) of 2,4-D.

(iv) Culture methods/conditions: Addition of Ficoll in the culture medium increased the density and the frequency of plant formation (Kao, 1981). Aerated and proper buffered medium along with high Ficoll and sucrose amount in liquid medium make anthers float, and sustain the microspores for a longer period of time under sugar – starvation. Wei *et al* (1986) indicated that sugar starvation led to induction of microspores and androgenesis.

Isolated Pollen Culture

Culture of isolated microspore (pollen) serves as ideal system providing large number of haploid single cells that can be exploited in mutant selection, genetic transformation, and further in studies pertaining to embryo development. Microspores can be isolated by two means: by making immature anthers to dehisce to release microspores into a suitable medium; the other mean involves disruption of anthers so as to expose microspores by homogenization or microblending, and then these are collected and cultured. Certain pre-treatments like sugar starvation, as in barley (Wei *et al.*, 1986), high temperature in *Brassica* (Chuong & Beversdorf, 1985) are known to trigger embryogenic process in microspores.

Ovule Culture

Haploids have also been successfully produced from cultured female gametophytes. Success in ovule culture was first observed with some Gymnosperms like *Zamia, Ephedra* and some Cycads. Other examples include barley, wheat and tobacco. Our understanding needs to be improved about (i) the events leading to induction of haploidy in female gametophytes, (ii) the factors regulating *in vitro* development of proembryo into the fully organized plants, and (iii) the differences in the growth patterns of *in vitro* development of unfertilized gametophytes (in ovules) and the male gametophytes (in pollen). The chief advantage of ovule culture seems to be that the majority of regenerants are green and are more genetically stable (Dunwell, 1985). San Noeum (1976) had the first success with *Hordeum vulgare* in obtaining haploids from cells of embryo sac, which later were made diploid using colchicine. Using cultured ovules, haploids have been produced in wheat, rice and maize. Keller (1986) observed that ovule culture for producing haploid may be better option than anther (microspore) culture for some taxonomic groups such as the Compositae. Following the success of San Noeum, Zhu & Wu (1979) obtained haploid plants from cultured unpollinated ovaries of *Triticum aestivum* and *Nicotiana tabacum*. Haploids have also been reported from unpollinated ovule cultures of *Gerbera jamesonii, Zea mays* and many other species (Yang & Zhou, 1990). The procedure for *in vitro* culture of unpollinated ovaries remains similar to that described for anther culture.

Chromosome Elimination

Instances are known where chromosome number was reduced to half in the somatic tissue either spontaneously or due to specific treatments, through a phenomenon called 'somatic reduction' or 'reductional mitosis'. A haploid branch on a watermelon was induced by irradiation of the seed. This could have occurred due to reduction of chromosome number in the somatic tissue, possibly through abnormalities in spindle organizer. Colchicine treatment is also found to cause reduction in chromosome number.

Barley haploids through crosses with *Hordeum bulbosum*

When interspecific hybrid of *H. vulgare* (Zn = 14) with *H. bulbosum* (Zn = 14) takes place, most of the progenies are barely haploids, remaining being diploid hybrids. Thus chromosomes of one parental genome are eliminated. Haploids in barley are produced after egg fertilization by progressive elimination of *H. bulbosum* chromosomes during early stages of embryo development.

Chromosome elimination is also a common feature among the *Hordeum* interspecific hybrids not involving *H. vulgare*, and has been reported in 19 such hybrids. e.g. *H. bulbosum* (4x) × *H. secalinum* (4x) gave haploids of *H. bulbosum*, while *H. secalinum* (4x) × *H. bulbosum* (2x and 4x), and *H. secalinum* (4x) × *H. vulgare* (2x) gave haploids of *H. secalinum* along with hybrids. Crosses between wheat × maize rescued embryos, as demonstrated by Bennett *et al.* As a result of elimination of maize chromosome, wheat haploids are obtained. The bulbosum technique offers several advantages compared to anther culture (Inagaki, 1987): (a) high frequency of haploid production (b) non-occurrence of albinos, and (c) immature haploid embryos are produced which act as material for haploid cell culture.

Mechanism of chromosome elimination

In *Hordeum* hybrids this is proposed to be due to (i) normal double fertilization taking place in interspecific crosses as evinced by cytological study, (ii) after fertilization, there is a gradual and selective elimination of *H. bulbosum* chromosome from nuclei of endosperms as well as embryo cells, leading to production of haploid embryos. Insufficient proteins in the developing embryo and endosperm, and the better ability of *vulgare* chromosomes to form spindle attachments compared to *bulbosum* chromosomes are the factors attributed for elimination of the *bulbosum* chromosomes.

Variation in chromosome structure in plants regenerated through anther culture: Variations such as dicentrics, telosomics, isochromosomes, deletions and breakages occur (De Buyser *et al.*, 1985). A number of pollen plants in wheat produced through tissue culture were mixoploids (Hu *et al.*, 1978). In root-tip cells and callus mitosis, evidence for aneuploid, mixoploid, chromosome breakage (through C-banding technique), endomitosis, and chromosome decentralization was noticed. Zeng & Ouyang (1980) observed that nuclear fusion and endomitosis double the chromosome number in wheat microspores *in vitro*. 'Anomalous cell wall' formation may account for ploidy variation in anther culture (Sun *et al.*, 1983). Mitotic abnormalities are observed during the late developmental stage of pollen callus, for instance, spindle fusion and nuclear endoreduplication producing polyploids. Whereas, chromosome bridge, multipolar mitosis, lagging chromosome and decentric chromosomes etc induce the formation of aneuploids.

Detection of Haploids

(1) Morphological basis: Also known as phenotypic effects. The vegetative and floral parts and the cell size in haploids are reduced relative to diploids. Size of stomata and pollen, and pollen abortion are other parameters to help initial screening.

(2) Use of genetic markers: The reason for the use of genetic markers is that hybrids and nonhybrids should differ; a failure of fertilization and parthenogenetic development of embryo can be traced in the progeny at the seed stage, seedling stage or adult plant stage. If the recognition needs to be done at the seed stage, then the marker should express itself in the late stage of embryo development (such as aleurone colour marker). If screening is to be done at seedling stage, then the marker should clearly express in the seedling (e.g. the purple seedling colour). This technique has been employed in developing maize monoploids.

Stability of Haploid Cultures

Usually haploid cultures are heterogeneous population of haploid, diploid and polyploid cells, and there is a continuous decrease in percentage of haploid cells in this mixture as the culture period increases. Somatic variation also occurs during long-term culture, and cultures become diploid. Such instability has been observed in *Atropa belladonna, Nicotiana tabacum* and *Azadirachta indica.* para-fluoro phenylalanine is being used to enhance haploid callus growth and maintenance of haploid nature.

Application of Haploids

Haploids have been very useful in agriculture and tree breeding. They are useful for two reasons: (1) the presence of one set of chromosomes facilitates the isolation of mutants, and (2) isogenic diploids can be obtained by chromosomes diplodization (chromosome doubling). Isogenic lines are those individuals that have the same genotype irrespective of their homo- or heterozygous nature. Through anther culture technique, haploids can be obtained in a few weeks and by doubling their chromosome number homozygous diploids can be produced in a single generation. Such fertile homozygous plants can be used for obtaining the inbred lines required for hybridization programme. These homozygous diploids retain their traits and remain stable. Anther culture helps solve the problems in genetic studies and plant breeding when conventional methods fail. Pollen haploids result in homozygosity and better expression of recessive characters, thereby augmenting the selection process. Recombinant gametes can be fully expressed at the level of homozygosity, and the recovery of recombinants as well as variants is possible through microsporoclonal variation. Since pollen plants derived from microspores are not gametes, it is called microsporoclonal variation. Their varied applications include:

(1) *Haploid production*: Androgenesis in tissue culture is the convenient and rapid method of production of large number of haploids in a number of taxa. Haploid production have also been achieved by parthenogenetic development of egg cells into ovule, chromosome deletion, delayed pollination, temperature shocks and regeneration from the callus. Pollen treated with X-rays have produced haploids in many crop plants including *Triticum* spp.

(2) *True Breeding lines*: The application of haploids in obtaining true breeding diploid lines by the use of chromosome doubling technique, finds its role in the production of new varieties of crops by anther culture. Isolation of multihaploids is possible, and regeneration of multisomic would help in the study of linkage relationships.

CONVENTIONAL BREEDING

Parent 1 x Parent 2
|
F1
|
F2
|
F3
|
|
F6
Homozygous lines produced

HAPLOID BREEDING

Parent 1 x Parent 2
|
F1(Anther culture)
|
Chromosome doubling
|
H1 (Homozygous diploid)
|
H2
Homozygous lines produced

(3) *Virus free plants*: Virus-free plants can be obtained through anther culture.

(4) *In Plant breeding*, haploids may be employed for use in studies related to experimental embryogenesis, plant breeding and cytogenetics, owing to the ease with which homozygous cultivars can be obtained in shorter times following diplodization by colchicine. Several double haploid lines have been selected for desired traits like improved yield, early flowering and increased disease resistance.

(5) *Mutation Breeding*: To induce mutation, young plantlets emerging from anthers are subjected to γ - radiation, and following selection mutants are easily selected. Chemical mutagens and x-rays are also be used to induce haploidy. By diplodization, stable diploids are produced which are of great economic importance in case of cereals, particularly, self-incompatible plants like rye. Monohaploids have the advantage for mutation breeders that the recessive mutations are immediately discernable. Mutation induction in haploid cells avoids chimeras formation (Nitsch *et al.*, 1969; Bajaj, 1983).

(6) *Vegetatively propagating crops*: Fertile diploid and tetraploids are produced in potato by anther culture, and tetraploids are also used for breeding purposes.

(7) Homozygosity is more important for plants which have a long juvenile phase (period from seed to flowering), such as bulbous plants, forestry tress and fruit trees.

(8) *Tree Breeding*: Long-term haploid callus cultures were first initiated from *Taxus brevifolia*. Androgenic haploids have been obtained in *Albizia*, *Populus*, *Camellia*, *Hevea*, *Prunus* and *Ulmus*.

(9) *Development of new varieties*: 'Mingo' – the first licensed cultivar of barley was produced using haploid induced *in vitro* based on chromosome elimination, bulbosum method (Kasha & Reinbergs, 1980). Pollen haploid

breeding has been exploited to develop a number of cultivars of wheat, maize, rice, pepper and fruits, having good agronomic charateristics like high yield, disease and drought resistance, good adaptation etc (Hu, 1995).

Problems associated with haploid induction/ production

(1) Often initial growth is followed by abortion of embryos.

(2) Regeneration of diploids and tetraploids at the same time as the haploids.

(3) Selective cell division, which must take place only in haploid microspores and not in other unwanted diploid tissues, is often impossible.

(4) Callus formation is usually detrimental.

(5) Isolation of a haploid from a mixture of haploids and higher ploidy levels is difficult since haploids are outgrown by stronger polyploids.

(6) Doubling of a haploid not necessarily yield a homozygote.

SUGGESTED READINGS

Bajaj, Y.P.S. (1983). *In: Handbook of Plant Cell Culture. Vol 1. Techniques for propagation and breeding*. MacMillan Publ. Co., New York, pp 228-287.

Bhojwani, S.S. & Bhatnagar, S.P. (1990). *Embryology of Angiosperms*. Vikas Publ. Co., New Delhi, pp 1-370.

Bhojwani, S.S. *et al.* (1986). *Plant tissue culture, a classified bibliography*. Elsevier, Amsterdam, pp 1-502.

Bourgin, J.P. & Nitsch, J.P. (1967). *Ann. Phsiol.* Veg., **9**, 377-382.

Chu, C.C., Hill, R.D. & Brule-Babel, A.L. (1990). *Plant Sci.*, **66**, 255-262.

Debergh, P.C. (1978). *Verh. kon. Acad. Wet. Belg.,* **40**, 1-70.

Dunwell (1985). *In: Cereal tissue and cell culture* (eds. Bight *et al.).* Martinum Nijhoff/W. Junk Publ., Dordrecht, pp 1-44.

De Buyser, J., Henry, Y. & Table, G. (1985). *Z. Pflanzenzucht.*, **95**, 23-34.

Guha, S. & Maheshwari, S.C. (1964). *Nature*, **204**, 497.

Hu, D. F. (1995). *Plant Cell Engineering Breeding Science and China*, pp 21-24.

Hu, H., His, T. & Chia, S. (1978). Acta Genet. Sin., **23**, 905-912.

Hunter, C.P. (1988). Ph.D. thesis, Wye College, University of London. Springer, Berlin.

Inagaki, M.N. (1987). *Jpn. J Breed.*, **37**, 275-282.

Kasha, K.J. & Reinbergs, E. (1980). *In: Achievements with haploids in barley research and breeding* (eds. Davies, D.R. & Hopwood, D.A.). The John Innes Charity, Norwich, pp 215-230.

Keller, W. A. (1986). Intl. Congr. Plant Tissue Cell Culture Abstr. 6, p. 120.

Murashige, T. & Skoog, F. (1962). *Physiol. Plant.*, **15**, 473-497.

Nitsch, C. (1974). *C.R. Acad. Sci., Paris*, **278**, 1031-1034.

Nitsch, J.P. & Nitsch, C. (1969). *Science*, **163**, 85-87.

Nitsch et al. (1969). *C.R. Acad. Sci.*, **269**, 1257-1278.

Ogawa, T. Hagio, T. & Ohkawa, Y. (1992). *Jpn. J Breed.*, **42**, 675-679.

San Noeum, L.H. (1976). *Ann. Amelior. Plantes*, **26**, 751-754.

Sun, C.S., Chu, C.C. & Li, H.Q. (1983). *Acta Bot. Sin*, **25**, 295-300.

Sunderland, N. (1978). *In: In vitro multiplication of woody species.* Round Table Conf., Belgium, Centre Rech. Agron. Etat., pp 65-86.

Tsay, S.S., Chen, J.J., Yeh, C.C. & Hsu, J.Y. (1988). *J Agric. Res. China*, **37**, 257-265.

Tulecke, W. (1953). *Science*, **117**, 599-600.

Wei, Z.M., Kyo, M. & Harada, H. (1986). *Theor. Appl. Genet.*, **72**, 252-255.

Wenzel & Foroughi-Wehr (1984). *In: Cell culture and somatic cell genetics. Vol 1. Laboratory precedures and their applications.* Orlando, Florida, pp 293-301.

White, P.R. (1934). *Plant Physiol.,* 9, 585-600.

Yang, H.Y. & Zhou, C. (1990). *In: Plant Tissue Culture: Applications and Limitations* (ed. Bhojwani, S.S.). Elsevier, Amsterdam, pp 240-258.

Zhu, Z.C. & Wu, H.S. (1979). *Acta Genet. Sin.,* **6**, 181-183.

QUESTIONS

1. Describe the procedures for obtaining haploids in plants.
2. Write short notes on the followings:

   (i) Haploid detection method

   (ii) Mechanism for haploid production

   (iii) Applications of haploids

   (iv) Androgenic haploids

CHAPTER - 7

MERISTEM CULTURE AND PRODUCTION OF DISEASE FREE PLANTS

Many plant species which propagate vegetatively are systematically infected by virus, fungi, bacteria and nematodes. In order to ensure high and sustainable yield, it is vital to produce disease-free stock plants. Unless large-scale population of pure inoculum of test pathogens are available, it is difficult to pursue the establishment of pathogenecity. Tissue culture techniques have made possible to carry out such experiments/tests in laboratory within short span of time.

Morel and Martin (1952) applied tissue culture techniques for elimination of viral infection in plants. They successfully obtained for the first time a virus-free plant (*Dahlia*) through shoot meristem culture. Several virus-free plants are regenerated *in vitro* such as *Pisum sativum, Trifolium repens* and *Citrus* sp., *Dactylis glomerata* (from mild mosaic mottle viruses) and *Lolium multiflorum* (from ryegrass mosaic virus) (Hu and Wang, 1983).

An increased resistance to leaf blight disease caused by *Phytophthora infestans* and *Alternaria solani* in clonal populations generated from mesophyll cell protoplasts of the potato cultivar - 'Russet Burbank' was demonstrated (Shepard *et al.*, 1980). Tissue culture techniques are employed for mass rearing of nematodes *in vitro* and screening of resistant breeding materials, nematicides and fungicides. Toxin resistant callus plants have also been obtained. Carlson (1973) obtained disease resistant plant by treating tissues to be cultivated with toxins, then by regenerating plants from the stable cell clones. Similarly, chlorosis resistant plants are regenerated after treating haploid tobacco cells with an analog to a bacterial toxin. Disease resistant plants induced by these techniques are sugarcane, clover, tobacco, corn, potato, etc. Some of the important reports in this regard are presented in the Table 7.1.

Leaves of potato regenerants, produced from calli and selected for resistance to culture filtrate of *P. infestans*, showed greater resistance to this filtrate than the leaves of control (Behnke, 1980). Different factors like temperature, inoculum density, phytochrome affect the expression of resistance in cultured plants. The expression of race specific resistance in tissue culture is demonstrated for potato in response to *P. infestans*. Besides, other systems in which expression of disease resistance in tissue cultures have been studied include tomato/*P. infestans* system, tobacco/*Pseudomonas* sp., soybean/*P. megasperma* var. sojal system, tobacco/T.M.V. system etc.

There are five methods employed for producing virus-free plants: (i) Meristem culture (ii) Heat treatment (iii) Heat treatment followed by meristem culture (iv) Adventitious shoot formation followed by meristem culture, and (v) Grafting of meristems on virus-free root stocks (also called 'micro-grafting'). Virus-free plants are sometimes also obtained by callus or protoplast culture. Meristem culture is often used to obtain plants free from bacteria or fungi.

Meristem Culture

Limasset and Cornuet (1949) observed that in systematically infected plants virus concentration decreased as they approached the apical meristem. In apical meristem, no virus was detectable in 50% of the materials tested. Accordingly, Morel and Martin (1952) suggested isolation of the apical meristem of a systematically infected plant *in vitro* to obtain a virus-free plant. They were successful in proving this hypothesis and obtained virus-free *Dahlia* plants through meristem culture of the infected plants. Meristem culture is also used extensively in germplasm preservation of endangered crop species.

The meristem is a dome of actively dividing cells, about 0.1 mm in diameter and 0.25 mm long. Meristem of both main shoot and auxillary buds are used for obtaining virus-free plants. Shoots of all angiosperms and gymnosperms grow by virtue of their apical meristems. The apical meristems are first formed during embryo development and they remain (except in dormant buds) in an active state of division throughout the vegetative phase of the plant. The totipotency of the plant cells form the basis of meristem culture technique.

In principle, all shoot meristems of a plant may act as starting material but the chance of success is dependent on the type of bud or shoot (terminal or axillary) and/or the position (basal or terminal) of the bud (Styer and Chin, 1983). After sterilization of shoot buds apical meristem (0.3 - 0.5 mm) is removed using forceps and scalpel. Isolation of larger meristems (with more leaf primordia) makes the chance of getting virus-free plants very small (Kartha and Gamborg, 1975). The isolated meristem is immediately transferred on to the surface of solid medium (which is preferred) and incubated in the culture room. In the media, usually low salt and high vitamins are used for the growth of isolated meristem. The pH lies between 5.4 and 6.0, and sugar concentration is 2-5% (w/v). The *in vitro* culture of meristem and shoot tip involves several phases such as initiation of culture and establishment of explant, growth and differentiation, proliferation of shoots and finally plantlet formation by rooting of shoots. All the conditions of nutrients, plant growth regulators and meristem are chosen empirically.

The meristem culture is influenced by factors like lateral position of the meristem as in *Chrysanthemum* sp., concentration and composition of plant growth regulators, pre-condition of the plants from which meristems are obtained, pH of the medium, and addition of gibberellic acid. Use of filter paper with liquid medium has been found better for growing meristem compared to static medium. The percentage of virus-free plants can also be dependent on season, as shown in case of carnation (van O's, 1964). Shoots of potato and carnation root best when meristem is isolated in the spring (Stone, 1963; Mellor & Stace-Smith, 1969). The low success (percentage) of isolated meristems that develop into virus-free plants could be due to infection, damage, drying out and browning.

Heat treatment

This method is particularly efficient against viruses and mycoplasmas attacking fruit trees, cassava and sugarcane (Kartha & Gamborg, 1975; Quak, 1977). Axillary buds in case of woody plants are made virus-free with heat teatment of 37-38°C. Similarly, virus-free grapevines have been obtained by treating shoots *in vitro* for 21 days at 35°C (Galzy, 1969).

Meristem culture and heat treatment

Prior treatment of plants with heat treatment increases the chance of obtaining more virus-free plants, specially when there is more than one virus present. Such treatment depends upon the precondition of plant, type of virus infection and its population, and the treatment itself. Experiments with *Chrysanthemum* cv. Blanche portivine supreme showed increase in virus-free plants from 9 to 90% by heat treatment of plant material on 10th or 30th day. Such treatment of heat is helpful in eliminating certain viruses which are difficult to be eliminated by normal meristem culture e.g. in potato infected with PVX, PVS. Plants are subjected to heat therapy prior to excision of meristem; by this way the virus concentration gets reduced and/or the virus-free zone is enlarged. The length of heat treatment (35-38°C) varies from 5-10 weeks. This procedure has been successfully employed with potatoes, carnation, strawberry and *Chrysanthemum*. Morel suggested the storage of potato tubers at 37-38°C for a month before taking up meristem culture.

Adventitious shoot formation followed by meristem culture

Mori *et al.* (1982) showed that adventitious shoots obtained *in vitro* from TMV infected leaf of tobacco were virus-free. This method of obtaining virus-free plants has also been successful with Lily (Allen, 1974) and hyacinth.

Micro-grafting

Using this technique, Murashige *et al.* (1972) and Navarro *et al.* (1975) eliminated two viruses while working with citrus. Later, this was also employed successfully with other plants like peach (Mosella, 1979), apple (Huang & Millikan, 1980) and *Eucalyptus* (Damiano *et al.*, 1986).

Regeneration of virus-free plants from callus and protoplasts

Cooper's experiment demonstrated that after a few sub-culturing tobacco callus can escape virus infection. Isolation of protoplasts from dark green zones of tobacco leaves infected by TMV produced virus-free plants.

Eradication of pathogens other than virus

Meristem culture can also get rid off bacteria (*Erwinia, Pseudomonas, Bacillus*) and fungi (*Fusarium, Rhizoctonia,*

Verticillium). Baker & Phillips (1962) obtained fungal-free plants of carnation by meristem culture - *Fusarium roseum f. cerialis* was eliminated, as a result. Theiler (1981) produced bacteria-free plants of *Pelargonium* which were initially infected with *Xanthomonas pelargonii.* Some mycoplasma-like bodies cause aster yellow disease in carrot. The repeated transferring of explants excised from these plants, in cultures eradicated this pathogen.

Mechanism of Virus elimination in meristems

(1) Virus inhibiting effect of plant growth regulators could be responsible for virus-free apical meristem, as apical meristem has higher concentration of plant growth regulators. This hypothesis has been ignored.

(2) Rapid cell division may lead to virus-free cells. Normal nucleo-proteins are synthesized first, and viral nucleoprotein is synthesized during cell elongation. So there is a competition between cell division and viral multiplication as evinced by experiments with tobacco callus.

Table 7.1. Examples of Meristem culture to eliminate virus

Plant Species	Virus Eliminated
Allium sativum	Mosaic virus
Brassica oleracea	Turnip mosaic virus, CaMV
Chrysanthemum sp.	Virus B
Dianthus barbatus	Mottle virus, Vein mottle virus, Ring spot virus
Ipomoea batata	Rugos mosaic virus
Petunia spp.	TMV
Saccharum officinarum	Mosaic virus
Solanum tuberosum	Potato Virus-X, Spindle tuber leaf roll virus
Musa spp	Cucumber mosaic virus

There is no general reason availabe to define the cause of virus elimination; this is due to the fact that some preferential viruses are only eliminated in meristem culture.

Effect of Virus Elimination

Following effects can be observed in virus-free plants raised through meristem culture:

(1) Increased growth and vigour of plants e.g. in rhubarb, cauliflower, *Chrysanthemum.*

(2) Improvement in rooting of stem cutting e.g. in *Pelargonium.*

(3) Increased flower size e.g. *Chrysanthemum, Narcissus.*

Virus Indexing

The availability of a rapid, effective and sensitive method for virus indexing is of paramount importance for success of a programme aimed at production of virus-free plants. All the plant materials need to be certified for virus-free status for initial exchange. Biological assays are used for virus indexing. e.g. sap of plants to test virus infection (sap transmission test). TEM serology or a combination of both are also used. Enzyme linked immunosorbent assay (ELISA) has been extensively used for virus indexing along with rDNA technology and nucleic acid hybridization techniques.

SUGGESTED READINGS

Allen (1974). *Acta Hort.*, **36**, 235-239.

Baker & Phillips (1962). *Phytopath.*, **52**, 1242-1244.

Behnke, M. (1980). *Theor. Appl. Genet.*, **56**, 151-152.

Carlson, P.S. (1973). *Science,* **180**, 1366-1368.

Damiano *et al.* (1986). *Hort. Sci.*, **21**, 684.

Galzy (1969). *Ann. Phytopathol. Physiol.*, **1**, 149-166.

Huang & Millikan (1980). *Hort. Sci.*, **15**, 741-743.

Hu, C. & Wang, P. (1983). *In: Handbook of Plant Cell Culture.* Vol 1. MacMillan Publ. Co., New York, pp 177-227.

Kartha & Gamborg (1975). *Phytopath.*, **65**, 826-828.

Limasset & Cornet (1949). *C.R. Acad. Sci.,* Paris, **228**, 1888-1890.

Mellor & Stace-Smith (1969). *Can. J. Bot.*, **47**, 1617-1621.

Morel, G. & Martin, C. (1952). *C.R. Acad. Sci.*, **235**, 1324-1325.

Mori *et al.* (1982). *Proc. 5th Intl. Congr. Plant Tissue Cell Culture,* **5**, 803-804.

Mosella (1979). *Diss. Universite Sciences Techn. Languedoc*, pp 1-202.

Murashige *et al.* (1972). *Hort. Sci.*, **7**, 118-119.

Navarro *et al.* (1975). *J. Am. Soc. Hort. Sci.*, **100**, 471-479.

Os, van (1964). *Neth. J. Plant Pathol.*, **70**, 18-26.

Quak (1977). *In vitro*, **13**, 194.

Shepard, J.F., Bidney, D. & Shahin, E. (1980). *Science,* **208**, 17-24.

Stone (1963). *Ann. Appl. Biol.*, **52**, 199-209.

Styer & Chin (1983). *Hortic. Rev.*, **5**, 221-277.

Theiler (1981). *Obstbau*, **6**, 97-103.

QUESTIONS

1. What is meristem culture? Describe its application in the production of disease-free plants.
2. Write short notes on the followings:

   (i) Mechanism of virus elimination

   (ii) Role of heat treatment in virus elimination from plants.

CHAPTER - 8

SOMACLONAL VARIATION

When plants are grown in culture, much genomic variability is produced. This variability can be found in callus and suspension cultures, tissue cultures, protoplast cultures and other *in vitro* culture systems. Such variation observed among plants regenerated from tissue cultures is called somaclonal variation (Scowcroft, 1985). Frequent genetic modifications are known to occur during the process of cell or tissue culture. These modifications tend to be inherited as mutations among the progeny of regenerated plants. The source of explant is a critical factor or variable for somaclonal variation. Plant cell, tissues and somatic tissues developed from various explants are sources for generating somaclonal variation. Somaclonal variation has been reported in a number of plant species such as sugarcane, potato, rice, wheat, *Brassica*, tobacco, tomato and others for various agronomic traits like disease resistance, plant height, maturity and for many physiological and biochemical traits (Table 8.1).

Factors determining the chance of mutation during *in vitro* culture

(1) The method of vegetative propagation used: Genetic stability is preserved if the *in vitro* propagation is carried out using single node or the axillary bud method. If by the use of growth regulators adventitious shoots are formed, then the probability of mutations

taking place increases. Further, method of production of these adventitious shoots also influence the mutation frequency. For instance, if a shoot arises from a single cell then chances of mutation are greater as compared to those arising from more cells (Norris *et al.*, 1983). Vasil (1986) indicated that probability of somaclonal variation is high with members of Gramineae when adventitious shoots arise in callus culture.

(2) The genotype used: Whereas some plants seldom exhibit the tendency to mutate after adventitious shoot formation (e.g. hyacinth, lily, etc.), others like *Begnonia* (Hakkaart & Versluijs, 1984) show high mutation rate. Some cultivars of plants like potato and oat mutate easily (Evans and Bravo, 1986). "Russet Burbank" cultivars of potato that have originated from leaf protoplasts were always mutated (Shepard *et al.*, 1980). Similar genetic instability was observed in *Nicotiana tabacum*, when plants were regenerated from protoplasts derived from cotyledon tissue (Dulieu and Barbier, 1982).

(3) Growth regulators: Some growth regulators such as 2,4-D, NAA and synthetic cytokinins lead to a high number of mutation. High amount of BA induced somaclonal variation in tobacco (Evans & Sharp, 1986). These have their effect on cell division, degree of deorganised growth and selective proliferation of cell types (Gould, 1984).

(4) Tissue types: The risk of mutation is high when the starting material happens to be differentiated tissue such as pith, as cell differentiation *in vivo* tends to produce polyploid cells (D'Amato, 1977). Polyploid shows chromosomal abnormalities more frequently than diploids (Scowcroft, 1985). The chance of mutation is less with undifferentiated tissue such as procambium, pericycle and cambium.

(5) Number of sub-culture: Repeated sub-culturing *in vitro* enhances the chance of mutation as observed for

protocorm culture of orchids, adventitious and axillary shoot formation, and for callus-, suspension- and single cell-cultures.

Reasons for Somaclonal Variation

Variations in somaclones may take place on account of various reasons (Bhaskaran, 1987):

(1) Pre-existing genetic variation in the explant tissue.

(2) Variants may be produced due to subtle changes because of gene mutations in cultures. Somaclonal variants for single recessive gene mutations are reported for maize, rice and wheat. Somatic crossing over followed by segregation may be responsible for homozygosity which in turn causes phenotypic expression of the recessive nature. Several gene mutations are reported in wheat somaclones (Larkin *et al.*, 1984).

(3) Numerical and structural changes in chromosomes during *in vitro* growth. Polyploidy, euploidy, inversion, deletion as well as translocation are reported to cause variation in regenerated plants (Lee & Phillips, 1988).

(4) Somatic crossing over involving symmetric and asymmetric recombination. Tissue culture may enhance the frequency of somatic crossing over.

(5) Changes in DNA, isoenzyme and protein profiles have been correlated with somaclonal variations in plants including potato, rice, barley and maize (Bajaj, 1990).

(6) Transposable elements have been reported to induce changes in strains of alfalfa and wheat (Groose & Bingham, 1984; Ahloowalia & Sherington, 1985).

(7) Intracellular mutagenic agents produced during *in vitro* growth.

(8) Molecular basis of somaclonal variation. Changes in chromosome number or structure, gene mutation, plasma gene mutation, alteration in gene expression,

gene amplification, meiotic crossing over - all are capable to lead to somaclonal variation. Further, somaclonal variation can be characterized based on morphological, isoenzymes and DNA markers like RAPDs, RFLPs, etc. (Jain, 1997a). Alteration of a single base pair results in changes of a single amino acid in polypetide sequence, thus producing somaclonal variants (Brettell *et al.*, 1986).

(9) Variation in DNA methylation could be responsible for tissue culture-induced mutagenesis (Phillips *et al.*, 1990). The DNA methylation and base sequence changes are frequent in maize callus and in regenerated plants (Brown *et al.*, 1991).

(10) Changes in organelle DNA. The classical example of such variation in tissue culture is reported in cytoplasmically controlled male sterility. A single mutation to male fertility and toxin insensitivity was due to frame shift mutation in mitochondrial DNA (Wise *et al.*, 1987).

(11) Epigenetic variation. Cultured cells upon exposure to stress factor result in altered expression of traits, and these changes are temporary and are not reflected in offsprings. Such epigenetic changes in tissue culture could be due to DNA amplification, DNA methylation or transposable elements. Phillips *et al.* (1994) suggested that somaclonal variation occurs by a stress-response mechanism.

Table 8.1. Examples of somaclonal variations in some crops

Crops	Traits	Inheritance	Reported by
Sugarcane	Resistance to *Helminthosporium sacchari*	Through vegetative propagation	Larkin & Scowcroft, 1983
Potato	Resistance to *Phytophthora infestans*	Through vegetative propagation	Shepard *et al.*, 1980

Tobacco	Resistance to *Pseudomonas* sp.	Sexual transmission	Thanutong *et al.*, 1983
	Leaf colour	Sexual transmission	Dulieu & Barbier, 1982
Tomato	Resistance to *Fusarium oxysporum*	Sexual transmission	Evans *et al.*, 1984
Wheat	Adh isozyme	Sexual transmission	Brettell *et al.*, 1986
Rice	Resistance to *Xanthomonas oryzae*	Sexual transmission	Ling *et al.*, 1985

Isolation of Somaclonal variants

It is grouped in two broad categories:

1. Screening, and 2. Cell selection

Screening

This involves the observation of large number of cells or regenerated plants for detecting variants; this is the only visible technique for isolation of mutants for yield and yield traits. This procedure is applied for isolating cell clones which produce high amount of certain biochemicals. Table 8.2 lists some of the examples where somaclonal variation is caused in various morphological traits.

Cell selection

In this case, a selection pressure is the basis of cell selection that allows survival or growth of desired variants only. e.g. selection of cells resistant to toxins, herbicides, high salt concentration etc. When the selection medium allows only the mutant cells to survive, it is termed as positive selection, whereas, in case of negative selection wild type cells divide and are killed by a counter selection agent like arsenate. The basis for selection of auxotroph mutants is the negative selection. Generally, two types of selection are applied: single step selection and multi-step selection. In the former one, the inhibitor is added to culture medium three times in excess compared to maximum inhibitory concentration (MIC) and cultures are maintained for several sub-culture regimes with the inhibitor. In a multi-step selection, a sub-lethal concentration (less than MIC) is added to culture medium. In the subsequent sub-cultures, a gradual increase in inhibitor level is maintained. This method has been employed in tobacco for herbicide tolerance (Miller and Hughes, 1980), and in maize for amino acid analogue resistance (Miao *et al.*, 1988).

Table 8.2. Somaclonal variation induced in morphological traits in some crops

Crops	Traits
Tobacco	Yield, plant height, leaf number, inflorescence type
Brassica	Abnormal leaves, slow growth, large pollens, multiple stem branching
Rice	Number of tillers per plant, frequency of fertile seeds, flowering period, plant height
Potato	Maturity period, tuber uniformity and skin colour
Sugarcane	Cane diameter, sugar yield, stalk number

Applications of Somaclonal variation

(1) Increased genetic variability for agronomic traits: Several useful somaclonal variants showing resistance to diseases, insects and tolerance to herbicides have been isolated (Table 8.3).

(2) *In vitro* selection: *In vitro* selection has been used to select agronomically desirable somaclones, including tolerance to pathotoxins, herbicides and diseases. Some examples include:

Table 8.3. Somaclonal variants having improved traits in some crop plants

Crops	Improved traits	Reported by
Wheat	Tolerance to drought stress	Sears *et al.*, 1992
	Resistance to barley yellow dwarf virus	Banks & Larkin, 1995
Rice	Lysine content	Sharp & Shaeffer, 1993
Brassica	Salt tolerance	Kirti *et al.*, 1991
Tobacco	Herbicide resistance	Chaleff & Ray, 1984
Potato	Resistance to *Fusarium* sp.	Behnke, 1980
Sugarcane	Resistance to Eyespot	Ramos Leal *et al.*, 1996

(a) Tobacco varients resistant to *Pseudomonas syringae* and *Alternaria alternata* have been obtained following selecting protoplast-derived callus on medium having pathotoxins (Thanutong *et al.*, 1983).

(b) Somaclonal variants with increased resistance to *Fusarium oxysporum* in celery has been obtained (Heath-Pagliuso *et al.*, 1989).

(c) Herbicide resistant variants have been obtained in cell culture (Chaleff & Mauvais, 1984; Chaleff & Parsons, 1978).

(3) 'Elite' Germplasm and Commercial cultivars: Variants with improved traits have been obtained/developed giving rise to new useful germplasm, as well as cultivars (Table 8.4).

Table 8.4. Examples of elite germplasm and commercial cultivars

Crops	Improved germplasm	Cultivars	Reported by
Tobacco	Resistance to potato virus Y		Chaplin *et al.*, 1980
	Resistance to blue mold		Rufty *et al.*, 1990

Sorghum	Resistance to acid soil		Duncan *et al.*, 1992
Celery	Resistance to *Fusarium* wilt		Heath-Pagliuso *et al.*, 1989
Red clover	Regeneration ability		Smith & Quesenberry, 1995
Rice		Dwarf, 10% higher yield, tolerance to salt	Winicov, 1996
Bell pepper		Fruits with fewer seeds	Evans, 1989

(4) Improved ornamental plants: Somaclonal variation is known to be employed commercially in ornamental plants (Duncan, 1997). A wide range of somaclonal variation for flower size, plant morphology, plant height in *Begoniax elatior* plants regenerated from callus have been reported (Jain, 1993a,b). Flower variation is reported in tissue culture - derived plants of carnation, *Chrysanthemum* and *Gerbera.*

Advantages and Drawbacks of Somaclonal variation

Advantages

(1) Cost - effectiveness.

(2) Source of genetic variability; helpful in crops with a narrow genetic base.

(3) Successfully removes defects in otherwise well-adapted cultivars.

(4) Improves various vegetatively and seed propagated species.

(5) Produces new variants.

Drawbacks

(1) Unstability of variants.

(2) Poor plant regeneration.

(3) Some somaclones possess undesirable features like sterility, etc.

(4) Variation is usually not new.

(5) Unpredictability associated with nature of somaclonal variation.

Gametoclonal Variation

Gametoclonal variation is defined as variation among plants regenerated from gametic cells in culture (Morrison & Evans, 1987). It is the result of changes that occur in the culture itself or genetic recombination which preceeds formation of gametes in an inbred F_1, F_2 or heterozygous individual. Such a variation may be subjected to selection at haploid level, and the chromosome number of selected plants can be doubled to obtain homozygous plants. The effects of variations are likely to be deleterious on agronomic performance. Some of the gametoclonal variation could be epigenetic as a result of DNA methylation in tissue culture, whereas other variants are due to single gene mutations and so persist. The main distinguishing points between somatic-derived somaclones and gametic-derived gametoclones are:

(1) Both dominant and recessive mutant genes induced by gametoclonal variation will express directly in haploid plants regenerated from microspores of diploid anthers, thereby helping in identification of new variants (gametoclones).

(2) Recombinants recovered in gametoclones are the outcome of meiotic crossing over.

(3) Gametoclone may be used after having stability by doubling its chromosome number.

Gametoclonal variations are more desirable compared to somaclonal variation because: (1) The mutant characteristic is expressed in the R_0 plants. i.e., plants regenerated from cells *in vitro*. (2) Cells having deleterious mutation may be expected to regenerate much less frequently in case of haploid than in diploid cells. Various examples of gametoclonal variations are given in the Table 8.5.

Table 8.5. Examples of Gametoclonal variations

Plants	Altered characteristics	Genetic basis
Hordeum vulgare	Plant height, yield, grain yield, culm length, heading date	Transgressive segregation, gametophytic selection
Brassica napus	Time to flower, leaf shape, pod size, flower type, glucinolate content	Mutation
Nicotiana sylvestris	Leaf colour, growth rate, DNA content	Gene amplification
N. tabacum	TMV resistance Heterochromatin content Days to flower, total alkaloids	Variation in segregation ratio Gene amplification Residual heterozygosity
Oryza sativa	Chlorophyll content, plant height.	Single genes induced in culture

SUGGESTED READINGS

Ahloowalia, B. S. & Sherington, J. (1985). *Euphytica*, **34**, 525-537.

Bajaj, Y.P.S. (1990). *Biotechnology in Agriculture Vol 11. Somaclonal variation in crop improvment* 1. Springer Verlag, Berlin, pp 1-685.

Banks, P.M. & Larkins, P.J. (1995). *Crop Sci.*, **35**, 599-600.

Behnke, M. (1980). *Z. Pflanzenzucht*, **85**, 254-258.

Bhaskaran, S. (1987). *In: Biotechnology in agriculture* (eds. Natesh, S., Chopra, V. L. & Ramachandran, S.). Oxford & IBH Publ. Co., New Delhi, pp 101-108.

Brettell, R. I. S., Dennis, E. S., Scowcroft, W. R. & Peacock, W. J. (1986). *Mol. Gen. Genet.*, **202**, 235-239.

Brown, P. T. H., Gobel, E. & Lorz, H. (1991). *Theor. Appl. Genet.*, **81**, 227-232.

Chaleff, R. & Mauvais, C. (1984). *Science*, **224**, 1443-1445.

Chaleff, R. & Parsons, N. (1978). *Proc. Natl. Acad. Sci., USA*, **75**, 5104-5107.

Chaleff, R. S. & Ray, T. B. (1984). *Science*, **223**, 1148-1151.

Chaplin, J. F., Burk, L. G., Gooding, G. V. & Powell, N. T. (1980). *Crop Sci.*, **20**, 677.

D' Amato, F. (1977). *In: Plant Cell, tissue and organ culture* (eds. Reinert & Bajaj). Springer Verlag, Berlin, pp 343-357.

Dulieu, H. & Barbier (1982). *In: Variability in Plants Regenerated from Tissue Culture* (eds. Earle, L. & Demarly, Y.). Praegar, New York, pp 211-229.

Duncan, R. R. (1997). *Adv. Agron.*, **58**, 201-240.

Duncan, R. R., Waskom, R. M., Miller, D. R., Hanning, G. E., Timm, D. A. & M. W. Nabors (1992). *Crop Sci.*, **32**, 1076-1077.

Evans, D. A. (1989). *Trends Genet.*, **5**, 46-50.

Evans & Bravo (1986). *In: Tissue culture as a plant production system for horticultural crops* (eds. Zimmerman *et al.*). M. Nijhoff, Dordrecht, pp 73-94.

Evans, D. A. & Sharp, W. R. (1986). *Bio/Tech.*, **4**, 528-532.

Evans, D. A., Sharp, W. R. & Madina-Filho, A. P. (1984). *Am. J. Bot.*, **71**, 759-774.

Gould, A. R. (1984). *CRC Crit. Rev. Plant Sci.*, **1**, 315-344.

Groose, R. W. & Bingham, E. T. (1984). *Crop Sci.*, **24**, 655-658.

Hakkaart & Versluijs (1984). *Vakbl. Bloem.*, **39**, 49-50.

Heath-Pagliuso, S., Pullman, J. & Rappaport, L. (1989). *Hort. Sci.*, **24**, 711-712.

Jain, S. M. (1993a). *Sci. Hort.*, **54**, 221-231.

Jain, S. M. (1993b). *In: Creating genetic variation in ornamentals* (eds. Schiva, T. & Mercuri, A.). Sanremo: Instituto Sperimentale per la Floriclotura, pp 93-103.

Jain, S. M., Saccardo, F., Rugini, E. & Grassoth, A. (1997). *Adv. Hort. Sci.*

Kirti, P. B., Hadi, S., Kumar, P. A. & Chopra, V. L. (1991). *Theor. Appl. Genet.*, **83**, 233-237.

Larkin, P. J. & Scowcroft, W. R. (1983). *Plant Cell Tissue Organ Cult.*, **2**, 111-122.

Larkin, P. J., Ryan, S. A., Brettell, R. I. S. & Scowcroft, W. R. (1984). *Theor. Appl. Genet.*, **67**, 443-455.

Lee, M. & Phillips, R. L. (1988). *Annu. Rev. Plant Physiol. Plant Mol Biol.*, **39**, 413-437.

Ling, D.H., Vidhyascharan, P., Borromeo, E.S., Zapata, F.J. & Mew, T.W. (1985). *Theor. Appl. Genet.*, **71**, 133-135.

Miao, S., Duncan, D. R. & Widholm, J.M. (1983). *Plant Cell Tiss. Org. Cult.*, **14**, 3-14.

Miller, O. K. & Hughes, K.W. (1980). *In vitro*, **16**, 1085-1091.

Morrison, R.A. & Evans, D.A. (1987). *Plant Breed. Rev.*, **5**, 359-391.

Norris *et al.* (1983). *Science*, **220**, 75-76.

Phillips, R. L., Kaeppler, S. M. & Peschke, V. M. (1990). *In: Proc. 7th Intl. Congr. Plant Tissue and Cell Culture*, Amsterdam, pp 131-141.

Phillips, R. L., Kaeppler, S. M. & Olhoft, P. (1994). *Proc. Natl. Acad. Sci.*, USA, **91**, 5222-5226.

Ramos Leal, R., Maribona, R. H., Ruiz, A. & Korneva, S. (1996). *Plant Breed.*, **115**, 3-42.

Rufty, R. C., Wernsman, E. A., Mani, C. E. & Gooding, Jr., G. V. (1990). *Crop Sci.*, **30**, 241-242.

Scowcroft, W. R. (1981). *In: Genetic flux in plants* (eds. Holm & Dennis). Springer Verlag, Vienna, pp 217-245.

Sears, R. G., Cox, T. S. & Paulsen, G. M. (1992). *Crop Sci.*, **32**, 507.

Sharp, F. T. & Schaeffer, G. W. (1993). *Plant Sci.*, **90**, 145-154.

Shepard, J.F., Bidney, D. & Shahin, E. (1980). *Science*, **208**, 17-24.

Smith, R. R. & Quesenberry, K. H. (1995). *Crop Sci.*, **35**, 295.

Thanutong, P., Furusawa, I. & Yamamoto, M. (1983). *Theor. Appl. Genet.*, **66**, 209-215.

Vasil, I.K. (1986). *In: Somaclonal variations and crop improvement* (ed. Semal, J.). M. Nijhoff, Dordrecht, pp 108-116.

Winicov, I. (1996). *Plant Sci.*, **13**, 105-111.

Wise, R. P., Pring, D. R. & Gengenbach, B. G. (1987). *Proc. Natl. Acad. Sci.*, USA, **84**, 2858-2862.

QUESTIONS

1. Define somaclonal variation. Discuss its significance in crop improvement.
2. Write short notes on the followings:
   (i) Causes of somaclonal variation
   (ii) Isolation strategies for somaclonal variants
   (iii) Applications of somaclonal variation

CHAPTER - 9

PRODUCTION OF STRESS TOLERANT PLANTS

Considering the negative effect of stresses on vegetation it would be profitable to adopt various biotechnological approaches to raise tolerant plants. Usually, conventional breeding programmes are employed to integrate favourable genes to induce stress tolerance in plants. The technique, however, is time consuming, cumbersome and more often unsuccessful. Tissue culture technique can be fruitfully utilized to induce tolerance which includes the use of some selective agents that permit the preferential survival and/or growth of a desired phenotype in an otherwise lethal environment.

Two types of selection methods are suggested (i) step-wise (gradual exposure), long term culture, (ii) shock treatment, short-term cultures. Selecting agents such as NaCl (for salt tolerance) or salts of heavy metals like Zn, Co, Cd, Hg, Al, Cu, etc. (for metal tolerance) etc. are added to the nutrient medium at different levels. Either the explants directly or the callus derived from them are sustained as mutants/variants and allowed to regenerate into complete plantlets. In gradual treatment, cultures are exposed in step-wise manner to the increasing levels of selecting agent. Whereas, in shock treatment, cultures are straight away subjected to a shock of high concentration, only those which could tolerate that level will survive. Suspension or protoplast cultures for mutant selection are considered better as they offer the finest single cell system. After every passage of 3 - 4 weeks on stress media, the cultures are monitored for their biomass, morphogenic potentiality, isozyme pattern and secondary metabolites production. Cultures showing best performance are selected as 'mutants'. Table 9.1 lists some important examples of *in vitro* raised stress tolerant plants.

Herbicide Resistant

Biochemical mechanisms exist in cultured cells which determine the resistance to biocide chemicals and provide the theoretical promise for selection *in vitro*. Zenk (1974) has reported cell suspension tolerant to 2,4-D when the cells were sub-cultured for six months in liquid medium supplemented with increasing amounts of herbicides. The cells were able to grow in 1 mM 2,4-D, while the control suspension was completely inhibited at 0.3 mM 2,4-D. The mechanism of tolerance involves an enhancement in the metabolism of 2,4-D. Widholm (1977) observed that carrot cells lost the tolerance to herbicides when transferred to 2,4-D free medium. It suggested that tolerance is the result of induction of enzymatic system responsible for the degradation of 2,4-D.

Chaleff and Persons (1978) reported for the first time that regenerated plants from stable cell lines were resistant to herbicides. They also obtained mutant tobacco plants with monogenetic dominant resistance to picloram. Selection for herbicide resistance *in vitro* has been done in *citrus*, barley and tomato.

Use of protoplasts as a system to select cell lines tolerant herbicides have not been extensively studied. Protoplast technology can be used to enhance the possibility to obtain monoclonal lines and offer the opportunity for intraspecific transfer of cytoplasmic factor of resistance to some type of herbicides.

Salt and Osmotic Tolerance

The first report on isolation of salt tolerant cell lines was by Zenk (1974) on the haploid cells of *Nicotiana sylvestris*. In order to raise stress tolerant plants through tissue cultures two strategies have been proposed: (i) longer (step-wise), (ii) shorter (shock treatment). Nabors *et al.* (1975) suggested tissue culture as a rapid method for producing NaCl-tolerant crop plants by isolating several lines of tobacco cells using longer stepwise selection method.

The most remarkable metabolic outcome of salt stress in higher plants is the rapid accumulation of proline (Delauney and Verma, 1993). Proline is known to accumulate under high levels of salinity and prevents damage from cellular dehydration by balancing the osmotic strength of the cytoplasm with that of the environment. However, when supplied exogenously, proline gives some osmotic tolerance to plants and cultured cells. Over production of proline increases root biomass and flower development in transgenic plants of tobacco under drought stress (Kavi Kishore *et al.*, 1995). The regenerated plants from salt tolerant calli of *Brassica juncea* indicate acquisi-tion of tolerance at the whole plant level (Jain *et al.*, 1990). Better performance of these somaclones in greenhouse, at higher levels of salt (0.5% NaCl) shows that acquired trait is genetically stable and transmitted to the progeny. Watad *et al.* (1985) obtained the stability of tolerant trait of *Nicotiana* cell lines by substituting the cytoplasm of *N. tabacuum* with *N. gossei*. The salt tolerant plants exposed to 300 mM NaCl were stable for 56 generations in absence of selection pressure. Likewise, Pandey and Ganapathy (1984) observed salt resistance for 12 weeks in *Cicer arietinum*. A cell line of brinjal (*Solanum melongena* L.) needed additional supply of 0.25% NaCl in the medium for optimum growth (Jain *et al.*, 1987). Gehlot *et al.* (1987) observed that in presence of NaCl after 8th or 9th subculture, cell lines of *Vigna aconitifolia* acted like halophytes as removal of NaCl caused death and decay of the cells. On the other hand, the requirements of *Bacopa* (Ali *et al.*, 1997) maintained *in vitro* for two years under salt stress survived successfully when transferred to field conditions. These reports thus suggest tissue culture technology as important screening system for salt tolerance.

Heavy Metal Tolerance

The possible mechanisms of metal tolerance in plants are:

(i) binding of metal to cell wall

(ii) reduced transportation through cell

(iii) compartmentation of ions in vacuoles

(iv) metal-metal resistant enzymes

(v) chelating by specific metal-binding proteins

Cell cultures are preferred in studying mechanism of metal tolerance (Miller, 1985), because the system provides more standardized experimental conditions and eliminate interferring processes of translocation (Reese *et al.*, 1986). Tissue culture technique has been used in selecting tomato cells tolerant to aluminum (Bennetzen and Adams, 1984), tobacco cells tolerant to cadmium (Reese and Roberts, 1984), *Linum* cells and plantlets tolerant to cadmium (Chakravarty and Srivastava, 1997). Through suspension culture, Ojima and Ohira (1983) selected two cell lines of carrot tolerant to Al and Mn by sub-culturing cells in excessive amount of $AlCl_3$ and $MnCl_3$ for several months.

Other stresses

Apart from salt and heavy metals, plants face several other environmental stresses like drought, frost, flood, high temperature, etc. The use of herbicides and pesticides in agriculture severely affect the crop plants. Appearance of new variants (somaclonal) in tissue cultures give scope of variability for better selection. The first report on mutant isolation through plant tissue cultures is on the selection of *Antirrhinum* suspension cells tolerant to high temperature.

In order to select drought tolerance in cells/plants PEG is used, which induces water stress in the cultures raised on a medium supplemented with PEG. Smith *et al.* (1985) suggested cell culture technology as effective mean for rapid evaluation and screening of plant genotypes for drought tolerance. Drought tolerant tomato plants are selected through PEG-induced water stress (Handa *et al.*, 1982, 1983).

Resistance to extremely high temperature is induced with suitable preconditioning treatments (Schroeder, 1963), under *in vitro* conditions. Dix and Street (1975) isolated chilling resistant cell lines of *N. sylvestris* and *Capsicum annuum* based on their ability to survive at 25°C after exposure for 21 days to -3°C and +5°C, respectively.

Table 9.1. Some examples of *in vitro* raised stress tolerant plants

Plants	Explants	Response Observed	Reported by
Medicago sativa		Salt tolerant cells require salt for plant growth	Black (1960)
Brassica juncea	Cotyledon	Salt tolerant plants	Jain *et al.* (1990)
Cicer areitinum	Callus	Stress tolerant callus selected	Pandey and Ganapathy (1984)
Cymbopogon meatinii	Callus	NaCl tolerant plants	—
Saccharum officinarum		Fe-efficient plants	—
Datura innoxia	Anther	Zn-tolerant plants	—
Datura innoxia	Haploid plants	NaCl-stress tolerant plants	Tyagi *et al.* (1981)
Oryza sativa		Salt stress tolerant calli	Subhashini and Reddy (1987)
Solanum nigrum	Callus	Salt tolerant plants	—

Cold-resistant cell lines from carrot suspension are have been isolated after treatment with EMS (Templeton-Somer *et al.* 1981). Hydroxyproline resistant lines of potato showed increased tolerance to both salinity and frost (Van Swaaij *et al.*, 1987). Studies are in progress to isolate and identify the tolerant genes and transfer them to desired species.

SUGGESTED READINGS

Ali, G., Purohit, M., Iqbal, M. & Srivastava, P.S. (1997). *Phytomorphology*, **47**, 97-106.

Bennetzen, J.L. & Adams, T.L. (1984). *Plant Cell Rep.*, **3**, 258-261.

Black, R. F. (1960). *Aust. J. Biol. Sci.*, **13**, 249.

Challeff, R. S. & Parsons, M.F. (1978). *Proc. Nat. Acad. Sci.*, USA, **75**, 5104-5107.

Chakravarty, B. & Srivastava, P.S. (1997). *Ann. Bot.*, **79**, 487-491.

Delauney, A. J. & Verma, D.P.S. (1993). *Plant J.*, **4**, 215-223.

Dix, P.J. & Street, H.E. (1975). *Plant Sci. Lett.*, **5**, 231-237.

Gehlot, H. S., Upadhyaya, A. & Sankhala, N. (1987). *In: Proc. Symp. Plant Cell and Tissue Culture of Economically Important Plants* (ed. Reddy, G.M.). Dept. of Genetics, Oamania University, Hyderabad, pp 135-140.

Handa, S., Bressan, R.A., Handa, A.K., Carpita, N.C. & Hasegawa, P.M. (1983). *Plant Physiol.*, **73**, 834-843.

Jain, R.K., Dhawan, R.S., Sharma, D.R. & Chowdhary, J.B. (1987). *In: Proc. Symp. Plant Cell and Tissue Culture of Economically Important Plants* (ed. Reddy, G.M.). Dept. of Genetics, Oamania University, Hyderabad, pp 155-162.

Jain, R.K., Jain, S., Nainawatee, H.S. & Chowdhary, J.B. (1990). *Euphytica*, **48**, 141-152.

Kavi Kishore, P.B., Hong, Z., Miao, G-H., Hu Chein-An A & Verma, D.P.S. (1995). *Plant Physiol.*, **108**, 1387-1394.

Miller, P. (1985). *Naturwissenschaften*, **72**, 664-665.

Nabors, M.W., Danial-Nadolny, L. & Brown, C. (1975). *Plant Sci. Lett.*, **4**, 155-159.

Ojima, K. & Ohira, K. (1983). *Plant Cell Rep.*, **24**, 789-797.

Pandey, R. & Ganapathy, P.S. (1984). *J. Exp. Bot.*, **35**, 1194-1199.

Reese, R.N. & Roberts, L.W. (1985). *J. Plant Physiol.*, **20**, 123-130.

Reese, R.N., McCall, R.D. & Roberts, L.W. (1986). *Environ. Exp. Bot.*, **26**, 169-173.

Schroeder, C.A. (1963). *Nature*, **200**, 1301-1302.

Smith, R.H., Bhaskaran, S. & Miller, F. R. (1985). *In vitro Cellular Dev. Biol.*, **21**, 541-545.

Subhashini, K. & Reddy, G.M. (1987). *In: Proc. Symp. Plant Cell and Tissue Culture of Economically Important Plants* (ed. Reddy, G.M.). Dept. of Genetics, Oamania University, Hyderabad.

Templeton-Sommers, K.M., Sharp, W.R. & Pfister, R.M. (1981). *Z. Pflanzenphysiol.*, **103**, 139-148.

Tyagi, A.K., Rashid, A. & Maheshwari, S.C. (1981). *Protoplasma*, **105**, 327-332.

Van Swaaij, A.C., Nijdam, H., Jacobsen, E. & Feenstra, W.J. (1987). *Euphytica*, **36**, 369-380.

Watad, A.A., Lerner, R.H. & Reinhold, L. (1985). *Physiologie Vegetale*, **23**, 887-894.

Zenk, M.H. (1974). *In: Haploids in higher plants* (ed. Kasha, K.J.). University of Guelph Press, Guelph, Canada, pp 339-353.

QUESTIONS

1. Write short notes on the followings:
   (i) Production of salt and osmotic stress tolerant plants through *in vitro* technology
   (ii) Production of herbicide tolerant plants *in vitro*.

CHAPTER - 10

CRYOPRESERVATION

The concept of cryopreservation is based on arrest of metabolic function of biological material stored at low temperature (-196°C). This ultra-low temperature is provided by the use of liquid nitrogen. Due to this treatment of the biological materials, almost all the metabolic activities of cells stop and the materials can be preserved for a long time. The importance of this technology was realized as a result of losses of these valuable biological materials during storage (Bajaj & Reinert, 1977). All conventional methods of storage failed to prevent losses caused by (i) climatic disturbances, (ii) attack of pathogens and pests (iii) natural disorders, and (iv) economic and political issues. The subsequent knowledge about the cryoprotectants like dimethyl sulfoxide (DMSO) and glycerol helped develop the effective protocol for cryopreservation (Dodds & Roberts, 1985). The earliest record of use of this technique dates back in 1958 when Sun attempted to preserve desiccated seedlings of *Pisum*. Since then, cryopreservation has been effectively employed for tissue culture of several plant species, for instance, potato, cassava, coconut, strawberry, sugarcane, wheat, etc. (Bajaj, 1987).

Methods of Cryopreservation

The entire process can be grouped into different steps:

(1) *Selection of materials/Need for cryopreservation*: The primary aim of cryopreservation is conserving valuable germplasm, particularly of those plants which are vegetatively propagated and which have short-lived seeds. The protocol in use should not create any genetic variability. Generally, young meristematic, cytoplasmic and small cells which are thin-walled and non-vacuolated are used for cryopreservation. Also, the cell density to be taken in the vials for cryopreservation should be high, to ensure greater survival rate.

(2) *Addition of cryoprotectors and the pre-treatment*: Cryoprotectors are chemicals which protect living cells against damage due to cryodestruction i.e., during freezing and thawing. The colligative properties of such chemicals reduce the harmful effect of electrolyte concentration which results from conversion of water into ice. These cryoprotectants can be categorised into:

(1) Permeating: e.g. glycerol, DMSO, methanol

(2) Non permeating: e.g. sugars, sugar alcohols, polyvinyl-pyrrolidone, polyethylene glycol (PEG), polyethylene oxide (PEO), hydroxystarch.

Besides, sucrose and some amino acids (e.g. proline) can also be used.

The cells need different pre-treatment periods with different compounds for proper cryoprotection. Use of higher concentration of these cryoprotectants should be discouraged as these are cytotoxic. DMSO at 5 - 10% and glycerol at 10 - 20% are preferred. A mixture of cryoprotectants is also used to improve the efficiency of the process (Bajaj, 1987); using a mixture of two or more cryoprotectants at low concentration is found to be more productive than a single cryoprotectant at a high concentration. The use of a mixture of cryoprotectants PEG, glucose and DMSO (10:8:10%) has been usefully demonstrated in sugarcane callus. The use of pretreatment or culture conditions before freezing of cells has beneficial effect for post-freezing survival of cells e.g.

sometimes, osmotically active compounds (osmoticum) like sucrose, sorbitol, mannitol are added to the culture medium to increase freezing resistance of the cells; these act by their dehydrating effect. Similarly, treatment with abscisic acid (in wheat suspension culture), and addition of amino acids like asparagine, proline and serine in culture medium have enhanced the survival rate of cells following cryopreservation.

(3) *Freezing*: Regulated cooling or pre-freezing is done to avoid any damage from intracellular crystal formation. Slow freezing, rapid freezing and droplet-freezing are the methods utilized. Slow-freezing prevents ice crystal formation intracellularly, by increasing dehydration of the cells. Rapid freezing is mostly used preserving shoot tips, as in case of strawberry, potato, somatic embryo of oil palms. Here, the cryoprotectant-treated specimens are directly soaked in liquid nitrogen i.e. at the temperature of -196°C; this temperature stops all metabolic activity and prevents biochemical injury. The cooling rate in rapid-freezing is so high, mostly several hundred degrees per minute. In case of droplet-freezing, materials/samples are frozen by slow cooling (0.5° C/min) to a temperature of -20 to -40°C, before putting them in liquid nitrogen.

(4) *Thawing, washing and regrowth*: Thawing is done by removing the specimen from frozen state (liquid nitrogen storage) and transferring them in a water bath at 35-40°C for about 1-2 min or until specimen is warmed up. This is followed by washing of plant specimen in order to remove toxic cryoprotectants. However, in cases where low or non-toxic cryoprotectants have been used, instead of washing culturing may be done. Following washing, cells are removed by centrifugation. It is possible that during storage in liquid nitrogen some cells die, and therefore, test of cell viability becomes pertinent step to be investigated. This can be achieved

by using technique of triphenyl tetrazolium chloride, fluorescein diacetate stain test, etc.

(5) *Regeneration of plantlets*: The viable cells are cultured on non-specific growth medium to finally get plantlets.

Applications of cryopreservation

Plant specimens in the form of cell suspensions, calli, somatic embryos, shoot tips, pollen grains, anthers, etc. have been subjected to cryopreservation and successfully plantlets have been regenerated (Bajaj, 1987). Some of the applications/advantages associated with cryopreservation include:

(1) Freeze storage of cell cultures (*Atropa, Glycine, Nicotiana,* somatic hybrid protoplasts of rice x pea) etc.

(2) Cryopreservation of rare genotypes and that of vegetatively propagated crops (e.g. potato, sugarcane etc.).

(3) Cryopreservation of recalcitrant seeds and embryos e.g. coconut, mango etc.

(4) Conservation of genetic uniformity.

(5) Cold acclimatization and frost resistance.

(6) Retention of morphogenetic potential.

(7) The technology save manpower, space, medium and there is no need of sub-culturing.

Germplasm Bank/Cell Cryo Bank/ Pollen Bank

Creation of germplasm bank and cell cryo bank have been suggested in order to preserve plant materials of high importance, in liquid nitrogen for a long-term storage. As a result of this, germplasm of recalcitrant seeds, endangered plants, plants having high value in medicine, agro-forestry etc. can be cryopreserved and stored usefully for fairly a longer period of time. Similarly, pollen grains can also be stored in frozen state, which would facilitate (i) hybridization between plants with flowering taking place at different times, (ii) reducing the spread of disease by pollination vectors, and (iii) greater longevity. Besides these, cryopreservation process has been made more effective by developing large sized cylinders of about 50 litres capacity where refilling of liquid nitrogen is not desired for about six months (Bajaj, 1987).

Several organizations like National Bureau of Plant Genetic Resources, New Delhi, Indian Council of Agricultural Research, New Delhi, Council of Scientific and Industrial Research, New Delhi, Department of Biotechnology, Govt. of India, New Delhi, International Rice Research Institute, Philippines, International Plant Germplasm Research Institute are engaged in preservation of plant germplasm based on this technology.

SUGGESTED READINGS

Bajaj, Y.P.S (1987). *In: Biotechnology in Agriculture* (eds. Natesh, S., Chopra, V.L. & Ramachandran, S.). Oxford & IBH Publishing Co., New Delhi, pp 109-131.

Bajaj, Y.P.S. & Reinert, J. (1977). *In: Applied and Fundamental Aspects of Plant Cell, Tissue and Organ Culture* (eds. Reinert, J. & Bajaj, Y.P.S.). Springer-Verlag, Berlin, pp 757-776.

Dodds, J.H. & Roberts, L.W. (1985). *Experiments in Plant Tissue Culture*, 2nd Ed. Cambridge University Press, Cambridge.

QUESTIONS

1. What is cryopreservation? Briefly describe procedure and advantages of cryopreservation.
2. Write short notes on the followings:

   (i) Cryoprotectors

   (ii) Cryo Bank

   (iii) Freezing and Thawing in cryopreservation.

CHAPTER - 11

BIODIVERSITY AND BIOTECHNOLOGY

Biodiversity became an important issue in the United Nations Conference on Environment and Development in June, 1992, and now it is discussed in non-technical sense. It refers to study of the distribution of the natural habitats of plants, animals and microbes and their climate conditions. The diversity of these natural habitats has expanded over a period of time through natural selection and also through introductions. Biodiversity includes diversity of form right from molecular level through individual, organisms, populations, communities, ecosystems, landscapes and biospheres. The United Nations Convention on Biological Diversity (CBD) defines biodiversity as "the variability among living organisms from all sources including terrestrial, marine and other aquatic ecosystems and the ecological complexes of which they are part". The term biodiversity encompasses ecological as well as evolutionary processes like:

(a) Natural Ecosystems: Forests, grasslands, wetlands, deserts, mountains, coastal and marine areas.

(b) Wild species and varieties: Species of plants, animals and microorganisms occurring in their natural state and the genetic variation within each of these species.

(c) Agricultural ecosystems: Farmlands, pastures, fisheries, aquaculture.

(d) Domesticated species and varieties: Species of crops, livestock, captive breeding of fish, and genetic variation within each of their species.

The biological diversity sustains the food, medicinal, clothing, shelter, spiritual, recreational and other needs of most of the population. Biodiversity is usually analyzed at three different levels:

(a) Diversity of biotic communities and ecosystems: Depending upon the abiotic resources and the conditions of the environment an ecosystem develops its own community of living organisms. e.g. forests, lakes, ponds, wetlands, grasslands etc are represented by diverse ecosystems having a distinct biotic community.

(b) Species diversity: This refers to diversity of species composition within a community.

(c) Genetic diversity: This refers to diversity of genetic organization within a species. A species having a large number of races, strains or varieties is considered to be more diverse in its genetic make up. Species and their populations contain a genetic library with number of genes ranging from about 1000 in bacteria to 4,00,000 or more in several flowering plants and a few animals. Genetic diversity provides potential for subsequent evolutionary change (Bawa *et al.*, 1991).

Biodiversity is described by two parameters:

(a) Point or α – diversity, which represents number of species in a specified area/region, and

(b) β – diversity, represented by turnover of species across space.

The former is less on remote islands and increases as we approach towards equator, whereas the latter usually depends on size of species range. Today, biodiversity has assumed enormous commercial importance and requires management on scientific lines. An understanding and identification of biodiversity, its conservation and characterization have all become crucial.

The world harbours an estimated over 10 million different types of organisms including plants, microorganisms and animals, out of which about 1.4 million species remain unrecorded. Around 3,00,000 spcies are that of plants including algae, fungi, mosses and higher plants (Myers *et al.*, 2000). About 4-40 million species on the earth are unknown and unmonitored. Other elements of biodiversity like genes, populations, communities and habitats too remain unidentified to a great extent. The agenda for Science, Environment and Development into the 21st Century (ASCEND-21), arising from a conference organised in 1991 in Vienna indicates the possibility of extinction to the extent of 20% of important species of plants, animals, fungi, and microorganisms, in the next 30 years. It becomes imperative that after having signed the Convention for biodiversity, the nations evolve their own biodiversity protocol and updates its regulations pertaining to biodiversity.

Table 11.1. Number of known plant groups from World and India

Plant Groups	Approximate Number	
	World	India
Fungi	1,00,000	12,480
Algae	14,000	1676
Bryophytes	25,000	2843
Fern	10,000	1012
Gymnosperms	900	64
Angiosperms	2,50,000	15,000
Monocots	50,000	
Dicots	2,00,000	

The number, kind and dimensions of plants and animals vary. Species may be present in various different forms ranging from age classes, sex to seasonal and ecological forms. Some organisms are unicellular and some are multicellular. Similarly, there occurs a variation in life cycle and reproductive behaviour among organisms. Table 11.1 and 11.2 list the diversity of living organisms in the world and/or India.

Table 11.2. Number of known species of animal groups in India

Animal Groups	Approximate Number
Non – Chordates	More than 70,000 spcies
Chordates	
Fishes	2500 species and sub – species
Amphibia	200 species
Reptilia	450 species
Birds	2000 species and sub – species
Mammals	850 species and sub - species

Of the estimated total global biodiversity of 10 million species in the world, India accounts for 5-6%. India is one of the twelve mega-diversity countries in the world. It's biodiversity richness is indicated by the large number of plant and animal species it possesses. The North-East region, Western Ghats, North-Western, Eastern Himalayas and Andaman & Nicobar Islands are home for rich floral and faunal diversity. The North-East India is one of the important megacentres of biodiversity. Diverse ecosystems with diverse range of biological diversity exists in this part of India. viz., Tropical wet evergreen, moist deciduous, sub-alpine, alpine forests, grasslands, fresh waters, rivers, swamp and marshy wet lands. Arunachal Pradesh, which is the largest state of NE India covering an area of 83,743 sq km, is one of the biodiversity hot spot areas in the world. The rich biodiversity in this state is due to its variation of altitude ranging from 150 m to 6500 m and varied climatic conditions. It has 4500 species of Angiosperms including 600 species of orchids, 500 species of medicinal plants, 52 species of *Rhododendron*, 18 species of *Hedychium*, 45 species of bamboo, 20 species of cane and 35 species of Gymnosperms. The forests of Arunachal Pradesh contain many rare and endangered medicinal plants like *Coptis teeta, Aconitum ferox, Rubia cordifolia, Oroxylum indicum, Piper mullesua, Illicium griffithii, Terminalia arjuna, Podophyllum hexandrum, Rhododendron arunachalensis* etc. These forests are also store house for many valuable trees like *Terminalia, Shorea, Mesua, Artocarpus, Albizia, Quercus* etc. About one fifth of the faunal wealth of India is estimated to be found in Arunachal Pradesh. The Namdapha Biosphere Reserve in Arunachal is home for more than 96 species of mammals, 25 amphibians, 75 fishes and many reptiles.

The Indo-Burma and Western Ghats/Srilanka have been identified as two hot spots of biodiversity in the Indian sub continent, due to the richness of biodiversity in these regions. There are 25 hot spots identified recently in the world for priority conservation of biodiversity (Myers *et al.*, 2000). The hottest hot spots are Madagascar, the Philippines, Brazil's Atlantic forest, the Caribbean, Indo-Burma, Western Ghats/Srilanka and Eastern arc, and coastal forests of Tanzania/Kenya. About 40 hot spots of Indian flora have been recognized in term of endemism they show (Nayar, 1996). Whereas the Indo-Burma hot spot is home for about 13,500 species of plants and 2185 species of vertebrates (of these 7000 plant species and 528 vertebrate species are endemic), the Western Ghat/Srilanka hot spot harbors 4780 species of plants and 1073 species of vertebrates including 2180 species of endemic plants and 355 species of endemic vertebrates (Myers *et al.*, 2000).

Importance of Biodiversity

Biodiversity is important for biosphere's health, stability and proper functioning. It has been part and parcel of human life since time immemorial. It is important to human race for various reasons:

(a) It serves as important natural resource. These bio-resources not only provide human beings food, medicines and clothing, but also mental and spiritual welfare (Fitter, 1987). About 3000 plant species are regarded as source of food.

(b) As valuable genetic resources: Wild plants are important genetic resources and used in crop breeding programmes. Around 21,000 plants with medicinal uses are listed by WHO, out of which only 5000 species have been tested for their potential source of new drugs. Global herbal market is estimated to be around US $62 billion, and expected to reach US $3 trillion by 2050. In India, there are about 8000 species of medicinal plants. The Indian export of herbal medicine/material is around 446 crores which can go up to 3000 crores by 2005.

(c) The forests and their products serve as source of livelihood and others.

(d) It acts as an instrument of maintaining a stable and healthy ecosystems; this ensures optimum utilization of abiotic resources.

(e) It has a great economic implication. The unspoiled nature of exotic ecosytems have the real economic value. e.g. each lion in an African National Park has an annual visitor-attraction value of US $ 27000. In a real sense, value of biodiversity determines not just the sense of its worth but the way it is used or abused. Conservation and sustainable use of biodiversity is ingrained in the Indian ethos and culture. Millions of people derive their daily livelihood from forests, rivers, grasslands and seas. The practices and beliefs of forest dwelling tribes, nomadic communities, fishers folk and farmers remain closely interlinked with the biodiversity they possess.

Biodiversity is also important in several less obvious ways:

(i) *Pollination*: This is vital for fruit and vegetable production. Bees, other insects, bats, birds pollinate about 75% of world's staple crops.

(ii) *Healthy soil communities*: The top thin layer of earth crust – about 15 cm deep contains rich organic materials, minerals, nutrients, microbes, insects, worms etc which increase the fertility of soil and help improve plant productivity. The health of soil thus is dependent on diversity of these organisms. These breakdown and recyle the organic matter, furnishing nutrients that help crops, pastures and forests grow.

(iii) *Natural pest control*: Natural parasites and predators serve as free 'pest control'.

(iv) Biodiversity renders its services through its role in photosynthesis, transpiration, nutrient cycling, climate regulation, air and water management and waste management. The economic value of 17 different ecological services biodiversity covering 16 biomes has been extrapolated to be in the range of US $16-54 trillion ($10^{12}$) per year (Costanza *et al.*, 1997).

Today, very large areas in the world that were previously covered by forests, water bodies and the like, have been taken over. Also, human activities have caused tremendous damage to biodiversity. Biodiversity built up by nature over billions of years is now under threat.

Causes for the loss of biodiversity

(1) Destruction of natural ecosystems: Loss of natural habitat is the greatest threat to the survival of wild species. Habitat destruction and fragmentation are presently recognised as the most serious cause of species extinction. Habitat fragmentation increases the extinction rate by (a) reducing population size (b) enhancing stochastic processes and genetic bottlenecks (c) reducing or eliminating keystone predators, and (d) favouring edge-effect. Modernization and human activities like clearing of land for agriculture, mining and industries, highway construction, dam building etc destroy the natural habitat of most animals and plants. These organisms are adapted to live and reproduce in a specific area/ habitat and can not survive when the same is destroyed. Wildlife habitats have suffered to a great extent by destruction of virgin forests by loggers and settlers. Similarly, marine organisms, coral reefs have all suffered as a result of pollution and over fished. The tropical rain forests are the main stay of diverse variety of biological diversity, and they are being destroyed at a more rapid rate than any other type of wild habitat. Wetlands are often world's most productive systems helpful in regulating water flow and eliminating silt, sediments and pollutants from moving waters. These are now being drained and dried for agricultural purpose or human settlement. In Australia, New Zealand, California about 90% wetland systems have been lost. Mangrove and Coral Reef ecosystems are also being severely affected.

(2) Over-exploitation: Excessive exploitation of economic and valuable species for commercial utility has led to depletion of large number of population of plants, animals and microbes. Exploitation of raw materials of medicinal importance by commercial establish-

ments like drug companies for trading has severely affected the availability of medicinally important plants; this is quite evident in North-East India as this region is store house for medicinal plants like *Coptis teeta, Aconitum, Rubia cordifolia, Rauwolfia serpentina, Oroxylum indicum, Piper mullesua, Podophyllum hexandrum, Illicium griffithii, Picrorrhiza kurroa* etc. Several important plants like orchids and Rhododendrons are being heavily exploited. Faunal losses too pose a serious concern: excessive harvesting of fish, molluscs, sea-cows and turtles have led to near – extinction of these species.

(3) Wildlife trade: This involves capture of animals for pets, zoo specimens, and research subjects for their fur, skin, tusk, meat etc. Such capturing of wild animals for commercial application has made many species endangered. For instance, many primates have become extinct due to illegal killing of mothers to capture their babies for zoos and pet dealers. Gorillas, chimpanzees and other primates are killed for their meat, which is sold in African markets.

(4) Overhunting: The primary reason for hunting has been food. Commercially wild animals are hunted for their products like skin, tusk, antlers, fur, meat, perfumes, cosmetics etc. Rhino is hunted for its horns, tiger for skin and bones, elephant for ivory, muskdeer for musk. Overhunting has brought numerous species on the verge of extinction. e.g. Caribbean manatee, Asiatic lion, dugong and many species of pheasants. Farmers and ranchers in North America have almost eliminated the red wolf and prairie dogs, whereas herders in Africa have reduced the population of Simian wolf.

(5) Introduction of exotic species and competition among species: Introduction of exotic species leads to competition between them and the native species for

food and space, and is a major threat to many plants and animals. On many islands, population of native birds, reptiles and mammals declined after the introduction of domestic animals. Similarly, introduction of rabbits and goats in Pacific region has caused destruction of habitats of several plants, birds and reptiles.

(6) Deforestation: Increased population, shifting cultivation, permanent agriculture, mining, hydroelectric etc cause deforestation, which is occurring at the rate of 17 million hectare per year. i.e., 1 acre per second, which is quite alarming. It is estimated that every day 47,000 hectare of forest are destroyed, over 16,000 hectare of land turned to desert, and between 100 – 300 species become extinct. According to the report of World Wide Institute, alomost half of the forest cover of about 3 billion hectare in the world were lost between 1980 and 1995. The tropical evergreen forests in the world, covering about 7% of the earth's land surface, contain atleast 50% of the important species found on the earth. About 48% of world's plant species occur in forests or near it, and so declining of forest cover at such an alarming rate is not a healthy indication for the environment and human kind. Temperate rain forests are also endangered ecosystems. Of the original 31 million forests nearly 18.7 millions have already been cleared.

(7) Environmental Pollution: Air pollution through Green house effect, Ozone depletion layer, change in precipitation rates, acid rain are destroying vegetation and productivity. Most of rivers, fresh water reservoirs are polluted due to release of industrial wastes, sewage, agricultural run-off, and these contaminate not only water but also the soil (Dwivedi, 2004).

(8) Natural calamities: Landslides, floods, droughts, earthquakes, storm and hurricanes, severely affect population of plants, animals and microbes, of both terrestrial and aquatic origin.

(9) Lack of adequate environmental education and effective legislation

Rate of Extinction

According to one estimate, 2 – 8% tropical forest species will be lost between 1990 – 2015. An estimated 23,000 plant species and sub- species and 4452 animal species are threatened globally (IUCN Red List 1990). In India, 15 – 20% (i.e., over 2500 species) of total vascular flora are threatened. Extinction of tree species can prove disastrous as about 300 species of insects and birds species depend on a single tree species. About 80% of the world's 1,00,000 tree species are in the tropics, and 10% of all tree species are now threatened. Use of Species-area relations and Endemics-area relationship help indicate about the rate of species extinction in given habitats/patches. Red data book gives the list of threatened plants or animals of any region. The IUCN has published two volumes of Red Data Book on biodiversity at global level, mentioning these species under different categories/state of extinction processes: vulnerable, endangered, rare, extinct and threatened. Some endangered biodiversity across the world is given below (Source: The Times of India, June 13, 1992, New Delhi edition):

- North America's 40% cropland has become a desert.
- More than 60% of the pacific North West Coastal forest cut down.
- Around 5335 square miles of rain forest destroyed in Brazil every year.
- 71% trees in Czech and Sloval Republic threatened by air pollution.

- Indonesia's coral reefs face threat.
- Philippines has lost 90% of its Coral Reefs.
- Australia's 23% range and croplands have turned to desert.
- India and Srilanka have suffered large scale depletion of rain forest.
- Congo's 60% rain forest is slated for clearing.

Conservation of Biodiversity

Conservation of biodiversity is now a priority for International efforts. The concern for its conservation is manifested in 1972 during UN conference on Human Environment at Stockholm, the UN conference on Environment and Devlopment held in Rio de Jenerio in 1992, the World conference on Conservation and Development in 1986 and other Consensus Building Exercise. There are several steps recommended and methods devised to check the deteriorating state of biodiversity:

(1) Undisturbed sites/land should not be used for urbanization/development as such developmental activities like creating townships and recreation centres, deforestation etc lead to loss of biodiversity.

(2) Birth rates in developing countries should be checked and sustainable, high yielding agricultural systems be developed so that sustainable utilization and preservation of biodiversity go hand in hand.

(3) Catalogues of genetic resources and natural biological inventories should be prepared so that threatened and endangered speceis can be protected against extinction. Multitaxa inventory provides a complete and wider understanding of site, region or global biodiversity (Stork & Samways, 1995).

(4) Over-exploitation of natural resources, emission of harmful gases and wastes should be checked.

(5) *In situ* and *ex situ* conservation strategies: **A biotechnological approach through *in vitro* conservation.**

(i) *In situ conservation*: This refers to (i) protection of habitat and ecosystem, (ii) conservation of plants and animals in their native habitat or even in man-made ecosystems where they naturally occur. e.g. National Parks, Sanctuaries, National Monuments, Nature Reserves, Cultural landscapes, Biosphere Reserves etc. This approach of conservation highlights protection of ecosystems for conservation of diversity of genes, populations, species, communities and various ecological processes. Networking of protected areas for high conservation strategy is the main stay of *in situ* conservation. The indigenous ethos of 'sacred groves' and `sacred lakes' set by local inhabitants, in fact, predate the modern concept of protected areas. There are 448 wildlife sanctuaries, 85 National Parks and 10 Biosphere Reserves in India, covering about 4.2% of total geographical area (MoEF Report, 1999). The Biosphere Reserve gives a framework for the study and conservation of the environment and for the sustainable use of natural resources. There are 411 such sites in 94 countries. *In situ* conservation facilitates evolution and ecological balance having indirect benefits such as watershed maintenance, wildlife habitat protection and environmental stabilization.

Table 11.3. Ten Biosphere Reserves in India

Biosphere Reserves	Approximate Area Covered (km^2)	State
Kaziranga	37,823	Assam
Great Nicobar	885	Andaman & Nicobar Islands
Gulf of Mannar	555	Tamil Nadu
Manas	2,837	Assam
Nanda Devi	2,600	Uttar Pradesh

Namdapha	7,000	Arunachal Pradesh
Nilgiri	5,520	Karnataka, Tamil Nadu, Kerala
Nokrek	160	Meghalaya
Sundarbans (Mangroves)	9,630	West Bengal
Uttarakhand	3,941	Uttar Pradesh

A drawback associated to *in situ* conservation is that it is prone to natural calamities such as floods, droughts, diseases and pests. Also, it requires large land space.

(ii) *Ex situ conservation*: This refers to conservation of samples of genetic diversity away from their field habitats, and done through botanic and zoological gardens, banks of germplasm, seed, pollen, seedlings, gene, DNA, tissue culture, and culture collections. There are 1023 gene banks across the world which conserve seed materials. In India, National Gene Bank was established in 1988 consisting of (i) cryo bank (ii) seed repository bank (iii) tissue culture repository bank. In 1986, the Department of Biotechnology, Govt. of India helped establish National Facility for Plant Tissue Culture Repository (NFPTCR) with the aim to develop suitable *in vitro* conservation technologies for medium and long-term conservation of micropropagated agri-horticultural and plantation crops. This centre also aims at cryopreservation of pollens, seeds and *in vitro* cultures. This provides stable long-term storage of plant genetic resources using methods of desiccation, slow freezing and vitrification. Several NGOs, Seed Savers Exchange and industries are working in tandem to protect biodiversity. **Tissue culture methods** help in multiplication of threatened and endangered plant species. Further, techniques like artificial insemination and embryo transfer are being used for multiplication of endangered animal species.

Crop based research centres of Consultative Groups on International Agricultural Research (CGIAR) support the *ex situ* conservation of crop genetic resources. In 1974, an International research centre exclusively devoted to plant germplasm research was established and named as International Bureau of Plant Germplasm Research; later it was named as International Plant Germplasm Research Institute in 1992.

During 1950s the scientists realised the mass scale replacement of land races by improved cultivars and the need to conserve them. This led to establishment of first National Seed Storage Laboratory of USDA in 1958 at Fort Colins. Subsequently several International Agricultural Research Centres were established and each crop-based centres built germplasm storage facilities.

Ex situ conservation using **tissue culture** offers huge advantage for conservation of vegetativley propagated plants, species with recalcitrant seeds, species with longer vegetative period prior to seed set. This is specially advantageous for species which are either asexually propagated or lack natural seed production mechanism, or can not be propagated uniformly. Unlike of *in situ* conservation, it requires little land space since a large number of cultures can be stored in a small space. For instance, according to one estimate, around 800 cultivars of grapes in six replicates that would require about 1 hactere field space, can be maintained in cultures in 2 m^2 of laboratory space. Tissue culture also provides a useful method of processing propagules outside the seed harvesting season. Various tissue culture techniques employed for *ex situ* conservation of biodiversity include meristem culture, shoot tiop culture, axillary buds, anther and embryo cultures, single cell suspension, protoplast cultures, callus, organ cultures, cryopreservation, synthetic seeds etc. The short - to medium - term storage aimed at delaying the period of subculture without any harmful effect to plant tissue, assist in plant conservation programmes.

(6) Biochemical Markers (Isozymes) and Molecular Markers: RFLPs and RAPDs help in monitoring of genetic stability in the natural population of plants. Samples characterised on the basis of DNA sequence rather than on phenotype can provide better coverage of existing diversity. The molecular markers technique assesses diversity at genomic level and identify populations or individuals with desirable characters and variations between the taxa, so as to decide the high priority genotype for conservation.

(7) The information technology aimed at distance learning and electronic networking, designed and targeted at plant biodiversity conservation programmes in India will help promote the understanding of the rich germplasm and execute international conservation activities. The computer- based storage and retrieval systems of information on plant genetic resources will improve our knowledge and develop technology packages for conservation, and also ensure exchange of information including databases.

(8) Eco-tourism/green tourism: Excessive and unabated loss of natural environment has led to a growing awareness of non-consumptive value of these shrinking resources. The concept of 'eco-tourism' is one of the examples of current pattern away from resource exploitation for trading. Nature tourism generates huge foreign exchange in countries like Kenya, Nepal and Costa Rica, with total world-wide income in developing countries estimated at US $ 2-12 billion. In India, eco-tourism holds a strong promise and future.

(9) Environmental Education: By creating awareness (educating) the people about the causes of environmental menace including biodiversity through outreach programmes and environmental education (EE) related events (Dwivedi, 2002) like (i) environmental campaigns (ii) holding public interactions/meetings; eco-clubs (iii)

organizing workshops (iv) involving local NGOs. The environmental campaigns in form of interaction with people both in rural and urban areas can help establish symbiotic partnership between human beings and the environment. EE must become an integral part of all individuals so as to establish environmentally conscious societies. There is a need of commissioning reviews from a multiplicity of stakeholders and translation of such documents into major Indian languages. Role of media and information technology (Environmental Journalism) through newspaper, Disks/CD-ROMS, electronic and sources of bio-information have a great bearing on biodiversity protection. During environmental education, emphasis should be given on:

(i) awareness about environmental deterioration and health risk assessment.

(ii) Increased public participation.

(iii) Village – initiated programmes.

(iv) Improved access to information.

(v) Cultural relationships with biodiversity.

(vi) Distributive justice and decentralization.

(vii) Inclusion of tribals/indigenous people during formulations of strategies.

(viii) Intellectual Property Rights and Patent vis-à-vis biodiversity.

(ix) Environmental biotechnology.

(x) Conservation strategies for sustainable biodiversity.

(xi) Importance and relevance of environmental economics.

All these will lead to better management of biodiversity. A knowledge-based society, focusing creativity and diversity, can enlarge human alternatives. Besides, the networking of science and technology institutions and enterprises of both the private and the public sectors at the national and international levels should be strengthened.

National Biodiversity Strategy and Action Plan (NBSAP)

The Ministry of Environment and Forests in 1999 prepared a National Policy and Macro-level Action strategy on biodiversity through a consultative process. This dealt with policies, critical gaps and strategies needed for conservation and sustainable utilization of biological diversity. Detailed action plans at sub-state, state, regional, and national levels are being prepared; this has been made possible through funding from Global Environment Facility (GEF). All these information on action plans at different levels in consolidated form brings out the national level action plan. The NBSAP project envisages the assessment and stocktaking of biological diversity-related information at various levels. The emphasis under this programme is on gender sensitive decentralized planning. Government agencies, non-governmental groups and village communities are contributing towards conservation of biodiversity.

Intellectual Property Rights and Patents vis-à-vis Biodiversity

Intellectual Property Rights (IPR) is a collective term applied to a number of different types of legal right granted by state authority. Ethically, IPR can be justified from the fact that a person has a right to profit from ideas that he or she created and developed. Tribal's traditional methods/ practices (Indigenous knowledge, IK) play an important role in conserving the biodiversity and forest resources. IPR facilitates technology transfer and help establish a system that promotes public disclosure of new information. It stems from certain kinds of intellectual efforts and ingenuity. It has been in focus in recent years and biotechnology is responsible for much of this trend. It has various types:

(i) Patent

(ii) Registered Design

(iii) Trade Mark

(iv) Plant variety

(v) Copyright

(vi) Trade Secret

The World Intellectual Property Organization based in Geneva administers the Paris Convention and all subsequent conventions. Following the signing of TRIPS (Trade related aspects of intellectual property rights), countries which were party to the agreement were directed to protect their plant varieties by means of patent legislation. The first major study of international patent protection for biotechnology was published in 1985 by the Organization for Economic Cooperation and Development (OECD). Before a patent is granted, following specific conditions are to be satisfied: (i) the invention must be new and should have utility (ii) it must be inventive (iii) it must be disclosed in a way which enables a person of normal skill to reproduce it. Processes, products, compositions and the new uses – all can be patented in biosciences. Protected from competition, the patent allows the inventor to exploit the invention commercially. The society benefits by such exploitation through better products and after the expiry of the patent (which is normally 20 years from the date of application), the technology becomes freely available to the society.

Plant varieties are protected through Plant Breeder's Rights (PBR) or Plant Variety Rights (PVR). Farmer's Rights is a concept developed and adopted in FAO. Rural Advancement Fund International 'Farmer Rights (RAFI) in 1989 suggested that a tax should be payable on commercial biological material derived from developing countries, but used in developed countries. Patenting plays important role in biotechnological research. Biotechnology deals with living organisms, and innovations are related to creation of new varieties and breeds of plants, animals or microorganisms by new methods of genetic engineering. Several milestones have been achieved till date that have brought better understanding and clarity in patenting in biotechnology. e.g. in 1977, the Budapest Convention agreed on deposition of microorganisms in lieu of their description for patenting purposes. In 1987, the first patent for a higher category of animal was granted for the 'oncomouse'. In order to ensure benefit to the nation in commercial utilization of biodiversity, patenting related to it by indigenous research must be encouraged. Today, the Department of Biotechnology, Govt. of India facilitates patenting of biotechnology research findings through screening the applications and also bearing their cost of patenting.

Some Web Sites on Biodiversity

www.abi.org	www.biodiversity.uno.edu	www.bco.ec.gc.ca
www.conbio.bio.uci.edu	www.consbio.org	www.iucn.org
www.redlist.org	www.gbif.org	www.idrc.ca/biodiversity
www.wri.org/biodiv.	www.nhm.ac.uk/science/consulting/bioddetail.html	
www.biodiversityproject.org	www.worldwildlife.org/ben/proceedings.htm	
www.bionet_us.org	www.biodiversity.nl	www.bsponline.org
www.cabs.conservation.org	www.biodiv.org	www.wipo.org

www.nal.usda.gov/bic/Biotech_Patents

www.fplc.edu/tfield/Iptutors.htm

www.pk2id.delhi.nic.in

www.patents.cnidr.org/access/acess.html

www.bl.uk/services/sris/faq.html

www.patente.com/geneabst.txt

www.law.cornell.edu

SUGGESTED READINGS

Bawa, K.S., Schaal, B., Solbrig, O.T., Stearns, S., Templeton, A. & Gabor, V. (1991). *In: From genes to ecosystems: a research agenda for biodiversity* (ed. Solbrig, O.T.). IUBS/ SCOPE/ UNESCO, Cambridge, Mass., USA, pp 15-36.

Costanza, R. *et al.* (1997). *Nature*, **387**, 253-260.

Dwivedi, P. (2002). *In: 2nd International Conference on Plants and Environmental Pollution.* NBRI, Lucknow, pp 98-99.

Dwivedi, P. (2004). *Environmental Pollution and Environmental Management.* Scientific Publishers (India), Jodhpur.

Fitter, R. (1987). '*Wildlife for Man*'. Excerpts. IUCN Special Report 18 (1-3), 8-9.

Gadgil, M. (1991). *Evol. Trends Plants*, **5**, 3-8.

May, R.M. (2002). *Curr Sci*, **82**, 1325-1331.

MoEF (1999). *National Policy and Macrolevel Action Strategies on Biodiversity.* Ministry of Environment & Forests, Govt. of India, New Delhi.

MoEF and Kalpvriksh, National Biodiversity Strategy and Action Plan: Guidelines and concept Notes (2000). Ministry of Environment & Forests, Govt. of India, and kalpvriksh, New Delhi/Pune.

Myers, N., Mittermeler, R.A., Mittermeler, C.G., da Fonseca, G.A. B. & Kents, J. (2000). *Nature*, **403**, 853-858.

Ramakrishnan, P. S. (1996). *Nat. Resour.*, **32**, 11-19.

Stork, N.E. & samways, M.J. (1995). *In: Global Biodiversity Assessment* (ed. Heywood, V. H.). Cambridge University Press, Cambridge, U.K., pp 453-544.

Nayar, M.P. (1996). *'Hotspots' of Endemic Plants of India, Nepal and Bhutan.* TBGRI, Thiruvananthapuram.

Stork, N.E. & Samways, M.J. (1995). *In: Global Biodiversity Assessment* (ed. Heywood, V.H.). Cambridge University Press, Cambridge, U.K., pp 453-544.

QUESTIONS

1. Describe the causes for the loss of global biodiversity.
2. Discuss the importance of biodiversity.
3. Describe various methods of conservation of biological diversity.
4. Write short notes on:
   (i) *Ex situ* conservation.
   (ii) Intellectual Property Rights.
   (iii) Biotechnological approach of biodiversity conservation.
   (iv) Role of environmental education in biodiversity management.
   (v) NBSAP.